BIRDS – EVOLUTION, BEHAVIOR AND ECOLOGY

BIRD BIOGRAPHIES

Birds – Evolution, Behavior and Ecology

Additional books and e-books in this series can be found on Nova's website under the Series tab.

BIRDS – EVOLUTION, BEHAVIOR AND ECOLOGY

BIRD BIOGRAPHIES

ALICE E. BALL

snova
New York

Library of Congress Cataloging-in-Publication Data

ISBN: 978-1-53616-460-2

Published by Nova Science Publishers, Inc. † New York

To My Friend
Elizabeth Jones
In Loving Acknowledgment of Her Untiring Aid,
Unwavering Faith, and Inspiring Criticism
This Book Is Dedicated

Contents

List of Illustrations

FOREWORD

John Burroughs, in his delightful essay called "Birds and Poets" says: "The very idea of a bird is a symbol and a suggestion to the poet. A bird seems to be at the top of the scale, so vehement and intense is his life—large brained, large lunged, hot, ecstatic, his frame charged with buoyancy and his heart with song. The beautiful vagabonds, endowed with every grace, masters of all climes, and knowing no bounds,—how many human aspirations are realized in their free, holiday-lives—and how many suggestions to the poet in their flight and song."[1]

Long before the place of birds in the great scheme of nature was understood, they made their appeal: first, to primitive man, who had curious superstitions and created beautiful myths concerning them; next, to poets and dreamers of ancient civilizations, who used them in allusions beautiful with Oriental imagery; to artists, who delighted in portraying symbolism; to later poets and lovers of beauty, who perceived deep truths and revelations of God; and to scientists, who saw back of the phenomena of nature the marvelous laws of God.

It is interesting to follow the effect birds have had upon the development of man. Though the religion of the early Egyptians was largely worship of the sun and moon, yet reverence for birds entered into their faith and their ritual. The swallow, the heron, the hawk, the vulture, x

---

[1] Used with the permission of the Houghton, Mifflin Co., the authorized publishers.

the goose, and the ibis were all held sacred. The people of Egypt with their belief in transmigration, imagined the swallow and the heron as possible abiding-places for their souls after death.

The Chinese and Japanese have had interesting conceptions regarding birds that have been both symbolic and poetic. In Japan, wild ducks, geese, cocks, herons, and cranes have been highly honored. The people have built torii gates, or entrances to their temples, as "bird-rests" or perches for their sacred fowl.

The Greek and Roman mythologies abound in allusions to bird-life. It was natural that the powerful eagle should be held sacred to Jupiter, the lordly peacock to Juno, the wise owl to Minerva, the repulsive vulture that haunted battlefields to Mars, the beautiful swan to Apollo, and the cooing dove to Venus.

The American Indians regarded birds with great reverence. Their bird-myths are full of beauty. To them the eagle and the raven were especially sacred.

The dove was a cherished symbol of early Christian writers and painters. The pelican, too, was revered; it was the mediæval symbol of charity. The red breast of the robin was thought to have been caused by a prick of a thorn in Christ's crown as the bird strove to "wrench one single thorn away." The red crossbill's beak was believed to have been twisted in its attempt to remove the iron nail from Christ's blood-stained hand.

Burroughs continues: "The very oldest poets, the towering antique bards, seem to make very little mention of the song-birds. They loved better the soaring, swooping birds of prey, the eagle, the ominous birds, the vultures, xi the storks and cranes, or the clamorous sea-birds and the screaming hawk. These suited better the rugged, warlike character of the times, and the simple, powerful souls of the singers themselves. Homer must have heard the twittering of the swallows, the cry of the plover, the voice of the turtle (dove), and the warble of the nightingale; but they were not adequate symbols to express what he felt or to adorn his theme. Æschylus saw in the eagle the 'dog of Jove,' and his verse cuts like a sword with such a conception.

"It is not because the old bards were less as poets, but that they were more as men. To strong, susceptible characters, the music of nature is not confined to sweet sounds. The defiant scream of the hawk circling aloft, the wild whinney of the loon, the whooping of the crane, the booming of the bittern, the loud trumpeting of the migratory geese sounding down out of the midnight sky, or the wild crooning of the flocks of gulls—are much more welcome in certain moods than any and all mere bird-melodies, in keeping as they are with the shaggy and untamed features of ocean and woods, and suggesting something like Richard Wagner music in the ornithological orchestra."

As the life of man grew less warlike and heroic, as the humbler fireside virtues were honored and the amenities of life were cultivated, it is true that poets sang of the gentler, more beautiful aspects of nature. Wordsworth wrote of the skylark, the cuckoo, and the throstle, Shelley and Shakespeare of the skylark, Keats of the nightingale and of goldfinches, Tennyson of the swallow and the throstle. They were, however, all deeply sensitive to the wilder phases of nature—to the scudding cloud, the dashing
xii spray of the ocean, the raving and moaning of the tempest. They saw, too, as have many later poets, a spiritual significance and an inspiration as truly great and ennobling as the conceptions of the older bards.

Numerous American poets have found spiritual help, comfort, and inspiration in birds. Frank Bolles felt the presence of God in the forest where the Oven-bird sings:

> "Pouring out his spirit's gladness
> Toward the Source of life and being."

Celia Thaxter mused on God's care of man and bird:

> "For are we not God's children both,
> Thou, little Sandpiper, and I?"

Serenity and joy came to Edna Dean Proctor:

"My heart beside the bluebird, sings
And folds serene its weary wings."

Edward Rowland Sill voiced human need in his poem:

SPRING TWILIGHT

"Surely thus to sing, Robin,
Thou must have in sight,
Beautiful skies behind the shower,
And dawn beyond the night.
Would thy faith were mine, Robin!
Then, though night were long
All its silent hours would melt
Their shadow into song."

Beautiful memories that soothed pain came to Helen Hunt Jackson at the mere shadow of a bird's wing across her darkened window. Bird-song bowed Lucy Larcom's heart in reverence:

"Then will the birds sing anthems: for the earth and sky and air
Will seem a great cathedral, filled with beings dear and fair;
And long processions, from the time that bluebird notes begin
Till gentians fade, through forest-aisles will still move out and in."

All who appreciate Bryant's great poem "To a Waterfowl" may see God, not only "flying over the hill with the bird," but as the unfailing guide of the human soul.

"He who, from zone to zone,
Guides through the boundless sky thy certain flight,
In the long way that I must tread alone,
Will lead my steps aright."

No more triumphant lines exist in literature than those in Browning's "Paracelsus" which express faith in God's guidance of man and bird:

"I go to prove my soul!
I see my way as birds their trackless way.
I shall arrive: what time, what circuit first,
I ask not: but unless God send his hail
Or blinding fireballs, sleet or stifling snow,
In some time, his good time, I shall arrive:
He guides me and the bird."

The poets of the past generations may have written much about birds, but it is quite probable that they possessed very little accurate information regarding the service they render to the world. Longfellow alone has bequeathed to us, in his beautiful "Birds of Killingworth," a plea for the preservation of birds because of their practical use to man as well as their æsthetic and spiritual value:

"Plato, anticipating the Reviewers,
From his Republic banished without pity
The Poets; in this town of yours,
You put to death, by means of a Committee,
The ballad-singers and the Troubadours,
The street musicians of the heavenly city,
The birds, who make sweet music for us all
In our dark hours, as David did for Saul.

"Think of your woods and orchards without birds!
Of empty nests that cling to boughs and beams
As in an idiot's brain remembered words
Hang empty 'mid the cobwebs of his dreams!
Will bleat of flocks or bellowing of herds
Make up for the lost music, when your teams
Drag home the stingy harvest, and no more
The feathered gleaners follow to your door?

"You call them thieves and pillagers; but know
They are the winged wardens of your farms,
Who from the cornfields drive the insidious foe,
And from your harvests keep a hundred harms."

During this past century, the period of scientific investigation, birds have received a large share of attention. The immortal pioneers in American Ornithology, Audubon, Wilson, and Nuttall have been followed by a host of scientists who have done work of distinction along various lines. They have described the birds of both fertile and arid regions, as well as far distant lands, such as Alaska and the tundra of the North. They have made complete and valuable collections, the most noted of which are in the National Museum of Washington and the American Museum of Natural History in New York. The latter xv contains famous Habitat Groups with beautiful backgrounds, painted by distinguished bird-artists.

Scientists have studied the anatomy of birds, their eggs, their nests, and nestlings; an army of field-men have been recording observations on migration, on the molt of birds, their songs and call-notes, their food habits, especially with relation to their economic importance. The work of the Biological Survey in the Department of Agriculture at Washington has been of incalculable value; the examination of the contents of birds' stomachs has given indisputable evidence of the relation the different species bear to insect-life and thus to vegetation. The bulletins published by the Department and the leaflets issued by the National Association of Audubon Societies have been enormous factors in the preservation of bird-life in the United States.

Dr. A. K. Fisher, Professor F. E. L. Beal, Dr. Sylvester D. Judd, Dr. C. Hart Merriam, Dr. Henry W. Henshaw, Dr. E. W. Nelson, Dr. T. S. Palmer, and Dr. Wells T. Cooke have done work of special distinction in the Biological Survey, Mr. William Brewster and Mr. E. H. Forbush in Massachusetts, and Dr. Frank Chapman in New York.

To Dr. Fisher I am especially indebted for the right to incorporate into this book extracts from the bulletins of the Biological Survey, and to Mr.

Forbush for permission to quote from his admirable book "Useful Birds and Their Protection."

It has been my purpose to give, not only a portrait and a description of the birds I have chosen for this volume, but a summing up of the beneficial and injurious habits of each, gained from the highest authorities obtainable. The book is intended for beginners, or for those xvi who long to know birds intimately and intelligently, and wish to belong to the great army of bird-students who are "doing their bit" to preserve the bird-life of our country.

PREFACE*

The purpose of this book is to give, not only a portrait and a description of the birds, but a summing up of the beneficial and injurious habits of each, gained from the highest authorities obtainable. The book is intended for those who long to know birds intimately and intelligently, and wish to belong to the great army of bird-students who are "doing their bit" to preserve the bird-life of our country.

---

* This is an edited, augmented and reformatted version of "Bird Biographies" by Alice E. Ball, illustrated by Robert Bruce Horsfall, originally published by Dodd, Mead and Company, dated 1923.

ACKNOWLEDGMENTS

In the "Foreword" of this book I express my grateful appreciation to Dr. A. K. Fisher and Mr. E. H. Forbush for permission to use extracts from published works. I wish to add my thanks to Dr. Charles Richmond and Mr. Joseph Riley of the National Museum of Washington, for their courtesy in furnishing me with bird-skins from the National Museum collections and a copy of the A. O. U. Check-list of 1910, used for the descriptions and ranges of the birds described in the text.

I am indebted to Dr. John M. Clarke, Director of the State Museum of the University of New York, for the permission to make selections from Eaton's "Birds of New York"; also to Dr. Francis H. Herrick, of Western Reserve University, and Dr. Alexander Wetmore, of the Biological Survey, for the right to quote from their publications.

The selections from John Burroughs, Thoreau, Frank Bolles, Dallas Lore Sharp, Florence Merriam, Olive Thorne Miller, Henry W. Longfellow, E. R. Sill, Celia Thaxter, Lucy Larcom, and Edna Dean Proctor, are used by permission of, and by special arrangement with, The Houghton Mifflin Co., the authorized publishers. Three selections from Wilson Flagg's "Birds of New England" are used by special arrangement with the Page Co. of Boston.

To the Courtesy of D. Appleton & Co. I am indebted for the right to quote one stanza of Bryant's "To a Waterfowl," dates and selections from Frank M. Chapman's "Birds of Eastern North America"; to G. P. Putman's

Sons for the use of three extracts from Dr. Herrick's "Home Life of Wild Birds," and to Charles Scribner's Sons for Henry van Dyke's rendering of the song sparrow's song. I acknowledge also with thanks my obligation to Mr. T. Gilbert Pearson, for his permission to use six color-plates of the National Association of Audubon Societies and to quote from the Educational Leaflets of the Society.

To my friends, Dallas Lore Sharp, Mrs. Sylvester D. Judd, and Miss Harriet E. Richards, I desire to express my deep appreciation of their suggestions and criticisms. I am indebted to Mr. James P. Chapin, Assistant-Curator at the American Museum of Natural History, New York, for a critical reading of the manuscript.

Part 1 - Descriptions and Biographies of Our Common Winter Birds

Chapter 1

INTRODUCTION: WINTER BIRDS

Most people are surprised to learn that about sixty species of birds may be seen in the north-central part of Eastern North America during the winter months. Many of us, if questioned, would affirm that sparrows, crows, and jays are the only winter birds to be found. If someone opens for us the door which leads out into the great bird-world, we may say, as did the writer of the old couplet:

> "I hearing get, who had but ears,
> And sight, who had but eyes before,"

and we may then find, even during the winter season, a surprising wealth of bird-life to enrich our own.

In spite of wings that will bear them immeasurable distances, birds seem to have unusual loyalty to their native haunts, and they stay in the North until hunger impels them to seek friendlier climes. Those that remain may be grouped according to the kind of food upon which they subsist during the winter: first, birds that eat animal food; second, birds that eat vegetable food; and third, 2 those that eat the eggs or young of insects on tree-trunks and branches, or chisel them from the wood.

To the first group belong six species of owls and eight species of hawks, eagles, crows, gulls, shrikes, and about eight species of ducks.

They feed on mice and other small rodents, on smaller birds and poultry, and on seafood such as fish, clams, mussels, and scallops.

The birds that live on vegetable food during the winter are numerous. Throughout the spring and summer months they may be useful destroyers of insects; but in winter they are able to subsist on what the woods and fields yield in the way of nuts, acorns, berries, and the seeds of grasses and weeds. Such are jays, red-headed woodpeckers, quail, grouse, and the following members of the finch or sparrow family: cardinals, pine grosbeaks, crossbills, goldfinches, snow buntings, juncos, tree sparrows, white-throated sparrows, redpolls, and pine siskins. Many of these are permanent residents, but juncos, snow buntings, tree sparrows, crossbills, pine grosbeaks, and a few others leave their homes in the far North when deep snows bury their food supply and resort to less severe climates. Winter wrens are found in some localities. A few robins, bluebirds, meadowlarks, and flickers, remain North during open winters.

The third group of winter birds consists of downy and hairy woodpeckers, chickadees, tufted titmice, brown creepers, nuthatches, and golden-crowned kinglets. They glean insect-eggs from the bark of trees as a large part of their winter food-supply and form an exceedingly important group. The enormous number of insect-eggs eaten by them every year is almost incalculable. Every 3 part of a tree—the trunk, the large branches, and small twigs—is scrutinized by these industrious members of the Life-Saving Army of our forests.

Dr. Frank Chapman recommends beginning the study of birds in the winter, while the trees are leafless and the birds comparatively few in number. People who spread tables for them are frequently surprised at the number of species they attract and at the pleasure they experience in the companionship of their interesting winter visitors.

Chapter 2

PERMANENT RESIDENTS AND WINTER VISITORS

BIRDS SEEN DURING THE WINTER NEAR NEW YORK CITY[2]

The class of birds called permanent residents includes species which are to be found throughout the year. Dr. Chapman states that comparatively few species of this group are permanent residents in the strictest use of the term. "The Bob-white, Ruffed Grouse, and several of the owls are doubtless literally permanent residents, but it is not probable that the Bluebirds, for example, found here during the winter are the same birds which nested with us in the summer. Doubtless our winter Bluebirds pass the summer farther north, while our summer Bluebirds 4 winter farther south, but as a species, the Bluebird is a permanent resident."

---

[2] The above lists of Winter Residents and Visitors near New York City is taken from Dr. Frank M. Chapman's pamphlet, "The Birds of the Vicinity of New York City," a reprint from the "American Museum Journal" of the American Museum of Natural History. The lists and dates are used with the permission of Dr. R. C. Murphy, Acting Director of the American Museum of Natural History, and of D. Appleton & Co., Dr. Chapman's authorized publishers.

PERMANENT RESIDENTS

- Bob-white
- Ruffed Grouse
- 8 species of Hawks
- Bald Eagle
- 5 species of Owls
- Hairy Woodpecker
- Downy Woodpecker
- Red-headed Woodpecker
- Flicker
- Meadowlark
- Blue Jay
- American Crow
- Fish Crow
- House Sparrow
- Purple Finch
- American Goldfinch
- Song Sparrow
- Cardinal
- Cedar Waxwing
- Carolina Wren
- White-breasted Nuthatch
- Tufted Titmouse
- Chickadee
- Robin
- Bluebird
- Starling

Winter residents or winter visitants are birds that breed farther north and move southward during the winter months to obtain food. They may arrive in the fall and remain until spring.

WINTER RESIDENTS AND VISITORS

- Horned Lark
- American or Red Crossbill
- White-winged Crossbill
- Pine Grosbeak
- Pine Siskin
- Redpoll
- Tree Sparrow
- White-throated Sparrow
- Northern Shrike
- Myrtle Warbler
- Winter Wren
- Brown Creeper
- Snowflake
- Junco
- Red-breasted Nuthatch
- Golden-crowned Kinglet
- A few in winter.
- Rare or irregular in winter.

Grebes, Loons, Auks, Cormorants, Snowy Owls, and several species of Gulls and Ducks may also be found during the winter months in the vicinity of New York City.

Chapter 3

DESCRIPTIONS AND BIOGRAPHIES

THE BLUE JAY

Crow Family—Corvidæ

Length: About 11½ inches; 1½ inch longer than the robin; tail, over 5 inches long.

General Appearance: A *crested* grayish-blue bird, with *bright blue wings and tail*, barred and tipped with black and white. In flight, the long tail is conspicuous; it resembles a pointed fan.

Male and Female: Grayish-blue above, grayish-white below, lighter on throat and belly. Head with a conspicuous crest; forehead black; bill long, strong, and black. A black band that extends back of the crest and encircles the throat is widest across the breast. Wings bright blue, barred with black; the white tips of some of the feathers form bands and patches of white.

Note: A harsh *yăh, yăh, yăh*, or *jay, jay, jay*, which Thoreau says is "a true winter sound, wholly without sentiment."[3]

Song: A pleasant, flute-like strain: *Pedunkle, pedunkle, parlez-vous*. There is a sort of jerkiness about his love-song, as though his throat was

---

[3] From "Notes on New England Birds," by Henry D. Thoreau.

unaccustomed to make agreeable sounds. Jays are able to produce many strange noises, and appear to enjoy using their power.

Habitat: Woodlands; those containing oaks and other nut-bearing trees preferred.

Nest: A rough basket of twigs, with a soft lining of root-fibers.

Range: Eastern North America. A permanent resident of south-central Canada and eastern United States, west to the Dakotas, Colorado, and central Texas.

Figure 1. Blue jay.

This brilliant, handsome blue-coat never "hides his light under a bushel"; his noisy *jay-jay* always proclaims his presence. He would at times be unendurable, except that he never remains long in one place; he is on the leap constantly, with a dash and an impudent assurance that is amusing.

He is the "bad boy" of the bird neighborhood, the terror of the small birds. They seem to have the same fear of him that children have of a great bully. He swoops down upon them, worries and frightens them, robs their nests, and brings to his own spoiled fledglings eggs and young as tidbits.

He is a devoted husband and father, who shows his best traits in his family circle. He reminds one of certain human beings who take excellent care of their own, but who are neither good neighbors nor desirable citizens. Occasionally, however, he has family differences. My sister tells of watching a jay bring twig after twig for nest-building to his mate, who was evidently in a bad mood. She would have none of them; she seized each twig and threw it away with a disagreeable *yăh, yăh*. After repeated attempts, he gave it up and both flew away. My sister never learned what occurred later.

The jay is an inveterate tease. He delights in annoying poor half-blind owls in the day-time, by pecking at them from unexpected quarters. An owl has been known to seize the Tormentor and speedily put an end to his existence.

The blue jay is a member of the same family to which the crow belongs, and while totally different in appearance, resembles him in his cleverness, his fearlessness, and his audacious insolence. Dr. Henshaw, formerly of 8 the Biological Survey in Washington, brings the following accusation against this bird:

> "The blue jay is of a dual nature. Cautious and silent in the vicinity of its nest, away from it, it is bold and noisy. Sly in the commission of mischief, it is ever ready to scream 'thief' at the slightest disturbance. As usual in such cases, its remarks are applicable to none more than itself, a fact neighboring nest-holders know to their sorrow, for during the breeding season the jay lays heavy toll upon the eggs and young of other

birds, and in doing so deprives us of the services of species more beneficial than itself."[4]

Mr. E. R. Kalmbach, also of the Biological Survey, says that in winter jays eat the eggs of the tent caterpillar, and the larvæ of the brown-tail moth, besides waste grain, and "mast,"—the name given to vegetable food such as acorns, beechnuts, chestnuts, hazelnuts, etc. It likes pecans and cultivated fruit in their season—two other points against the bird. The wild fruits it selects are of no economic value.

Mr. Kalmbach concludes:

"The blue jay probably renders its best service to man in destroying grasshoppers late in the season and in feeding on hibernating insects and their eggs, as they do in the case of the tent caterpillar and brown-tail moth. Beetles and weevils of various kinds also fall as their prey. The severest criticism against the species is the destruction of other birds and their eggs. Where we wish to attract the latter in large numbers about our dooryards, in our parks, and in game 9 preserves, it will be well not to allow the jay to become too abundant."[5]

Wilson Flagg says:

"The blue jay is a true American. He is known throughout the continent and never visits any other country. At no season is he absent from our woods.

"He has a beautiful outward appearance, under which he conceals an unamiable temper and a propensity to mischief. There is no bird in our forest that is arrayed in equal splendor. But with all his beauty, he has, like the peacock, a harsh voice. He is a sort of Ishmael among the feathered tribes, who are startled at the sound of his voice and fear him as a bandit. There is no music in his nature; he is fit only for 'stratagems and spoils.'

"He is an industrious consumer of the larger insects and grubs, atoning in this way for some of his evil deeds. I cannot say, therefore, that

---

[4] Farmers' Bulletin 513, U. S. Department of Agriculture, Bureau of Biological Survey.
[5] Farmers' Bulletin 630, Biological Survey, U. S. Department of Agriculture.

I would consent to his banishment, for he is one of the most cheering tenants of the grove at a season when they have but few inhabitants."[6]

FLORIDA JAYS

Two species of jays are found in Florida. One, called the Florida blue jay, resembles its northern relative, except that it is somewhat smaller (10½ inches), is less brilliant in color, and has narrower, less conspicuous white tips to its feathers. These jays frequent live-oak trees. A flock of six or eight on the ground searching for acorns, is pleasing to the eye, but not to the ear.

A second species is called the Florida jay. The top and sides of its head are a *grayish-blue*; its neck, wings, 10 and tail are a *brighter blue*; *its back is a grayish-brown*; its under parts are gray, washed with brown, and faintly streaked on throat and breast. Its breast-band is *bluish*. This jay is found chiefly along the southern coast of Florida. The *absence of a crest* is its most distinguishing mark.

PACIFIC COAST JAYS

Two species of jays are common in California and its neighboring states. One, the Steller jay, enjoys a good reputation. It differs from its better-known relatives in appearance, also. Its *head*, *crest*, *throat*, *breast*, and *back* are a *brownish-black*; its *belly* and *rump* are *light blue*, its wings and tail purplish-blue, barred with black.

It is a shy bird and does not often approach the haunts of man. Its food is very like that of other jays, but its habits bring no condemnation upon it.[7]

The California jay is similar to the Florida Jay and may be easily distinguished by its blue head *without a crest*, its blue neck, wings, and

---

[6] From "Birds of New England," by Wilson Flagg.

[7] Farmers' Bulletin 630, Biological Survey, U. S. Department of Agriculture.

tail, its brown back, white throat, and gray under parts. This jay is a decided reprobate. Professor Beal has characterized it as follows:

> "It freely visits the stockyards near ranch buildings, and orchards and gardens. As a fruit stealer it is notorious. One instance is recorded where seven jays were shot from a prune tree, one after the other, the dead bodies being left under the tree until all were killed. So eager were the birds to get the fruit that the report of the gun and the sight of their dead did not deter them from coming 11 to the tree. In orchards, in canyons, or on hillsides adjacent to chaparral or other cover, great mischief is done by this bird. In one such case an orchard was under observation at a time when the prune crop was ripening, and jays in a continuous stream were seen to come down a small ravine to the orchard, prey upon the fruit, and return.
>
> "Fruit stealing, however, is only one of the sins of the California jay. That it robs hens' nests is universal testimony. A case is reported of a hen having a nest under a clump of bushes; every day a jay came to a tree a few rods away, and when it heard the cackle of the hen announcing a new egg it flew at once to the nest. At the same time the mistress of the house hastened to the spot to secure the prize, but in most cases the jay won the race. This is only one of many similar cases recounted. The jays have learned just what the cackle of the hen means. Another case more serious is that related by a man engaged in raising white leghorn fowls on a ranch several miles from a canyon. He stated that when the chicks were very young the jays attacked and killed them by a few blows of the beak and then pecked open the skull and ate out the brains. In spite of all efforts to protect the chicks and kill the jays, the losses in this way were serious."[8]

THE CANADA JAY

The Canada jay is similar in form and size to its blue relatives, but has the coloring of a northern winter landscape—gray, black, and white. This jay has no crest; 12 the back of its head and nape are black; the forehead

[8] Farmers' Bulletin 630, Biological Survey, U. S. Department of Agriculture.

and neck are white; the upper parts are gray, with darker gray wings and tail; under parts, light gray; tail, long; plumage, fluffy and fur-like.

This bird is found in the forests of Canada and in the northern part of the United States, where it is most common in the coniferous forests of Maine and Minnesota, in the wilder parts of the White and Green Mts., and in the Adirondacks.

Major Charles Bendire, in his interesting "Life Histories of American Birds," published by our government, writes the following amusing account of the Canada jay:

> "No bird is better known to the lumbermen, trappers and hunters along our northern border than the Canada Jay, which is a constant attendant at their camps, and affords them no little amusement during the lonely hours spent in the woods. To one not familiar with these birds it is astonishing how tame they become.
>
> "Mr. Manly Hardy writes: 'The Canada Jay is a constant resident of northern Maine, but in some seasons they are far more abundant than in others, being usually found in companies of from three to ten. They are the boldest of all our birds, except the Chickadee, and in cool impudence far surpass all others. They will enter tents, and often alight on the bow of a canoe where the paddle at every stroke comes within 18 inches of them. I know of nothing which can be eaten that they will not take, and I had one steal all my candles, pulling them out endwise one by one from a piece of birch bark they were rolled in, and another pecked a large hole in a cake of castile soap. A duck which I had picked and laid down for a few minutes had the entire breast eaten out by 13 one or more of these birds. I have seen one alight in the middle of my canoe and peck away at the carcass of a beaver I had skinned. They often spoil deer saddles by pecking into them near the kidneys. They do great damage to the trappers by stealing the bait from traps set for martens and minks and by eating trapped game; they will spoil a marten in a short time. They will sit quietly and see you build a log trap and bait it, and then, almost before your back is turned, you hear their hateful *ca-ca-ca* as they glide down and peer into it. They will work steadily carrying off meat and hiding it.'"

THE AMERICAN CROW

Crow Family—Corvidæ

The American crow is too well-known to need a description—merely a reference to the steely-blue or dark purple sheen of his "crow-black" plumage, and to the remarkable power of his long (twelve-inch) wings, which in flight show feather finger-tips at their ends.

One cannot but admire his strength and his absolute fearlessness, nor fail to be amused at his cleverness and his insolent bravado. Two or three crows, cawing hoarsely, will people a woodland in winter; while a flock, winging its way to the naked March woods, will cause a thrill of joy and expectancy, in spite of the knowledge that the advent of these black marauders means eternal vigilance to long-suffering farmers.

Dr. Sylvester D. Judd at Marshall Hall, Maryland, made an exhaustive study of the crow's food habits. He reported the following:

> "The crow is by all means the worst pilferer of the cornfield. Every year at Marshall Hall, as elsewhere, a part of the field must be replanted because of his 'pickings and stealings.' In 1899, the replanting was more extensive than usual—46 per cent of the 3½ bushels originally planted. This unusual ratio was probably caused by the failure of the cherry crop, which left the crow short of food."

Dr. Judd told of the "protective devices of tarring corn," which did not prevent the crows from pulling up 15 the grain in large quantities, though they did not eat it. He continued:

> "The injury to corn at other seasons than sprouting time is, as a general thing, comparatively insignificant, but in some years it has been important when the ears were in the milk. They then tear open the ears, and pick out the kernels in rapid succession. In the National Zoölogical Park at Washington during the summer of 1896, their depredations on an acre of corn were watched, and 50 per cent of the crop was found to have been ruined.

"The only scarecrows that proved effective at Marshall Hall were dead crows, and strings stretched on poles around the field and hung with long white streamers. Although in fall the number of marauders is greatly increased by reinforcements from the North, ripe corn sustains less injury from crows than roasting ears. One reason is the abundance of fall fruit.

"Wheat suffers comparatively little. When it is ripening, cherries and sprouting corn divert the crow's attention. After it is cut and gathered into the shock, however, they often join the English sparrows in removing the kernels. Oats are injured even less than wheat, though crows have been noticed feeding on them at harvest time."

While the crow is considered the arch-criminal of the bird-world, Dr. Judd ascribed to him a good habit—that of the dissemination of wild seeds in an unusual manner. He wrote:

"In November, 1899, a large flock on the wing was noticed in the distance, at a point opposite Fort Washington, several miles above Marshall Hall. They came on down the river in a line that at times stretched almost from one bank to the other. They circled several 16 times and alighted on the shore. The flock numbered at least a thousand, and hoarse caws and croaks gave evidence that it was made up to some extent of fish crows.

"After the birds had remained on shore about fifteen minutes, they were put to flight by a farmer's boy and flew on down the river. Going to the place where they had alighted, I found the sandy beach cut up for more than a hundred yards with their tracks. Many led out to the water, and floating black feathers here and there showed where baths had been taken.

"The most interesting trace of their sojourn, however, was several hundred pellets of fruit material, which they had ejected through their mouths and dropped on the ground. These pellets were about an inch in length and half an inch in diameter. They were of a deep purplish color, due to the fruit of woodbine, wild grape, and pokeberry, of which they were mainly composed. In 50 pellets collected there were only 11 seeds of other plants—namely, holly, bitter-sweet, and poison ivy. Pokeberry seeds were by far the most numerous. Mr. A. J. Pieters, of the Botanical

Division of the Department of Agriculture, germinated some of them, thus demonstrating the fact that they were distributed uninjured.

"The pellets were made up not only of seeds and skins, but largely of fruit pulp in an undigested state. It seems strange that the birds should have rid themselves of a substance that still contained a good deal of nutriment.

"Little is known of the distribution of fruit seeds by crows during migration, but it is certain that they do this work effectively while they fly to and from the roosts where they congregate in winter, for their feeding grounds often cover an area stretching out on all sides from the roosts 17 for 50 miles or more. It appears highly probable that the crows which are found in winter at Marshall Hall roost at Woodbridge, D. C., some 15 miles distant. There, in the midst of several acres of woodland, a crow dormitory is established, in which probably 100,000 crows sleep every winter night. It was visited in February, 1901, and the ground was found to be strewn with disgorged pellets."[9]

The fish crow (16 inches long) is three inches smaller than the common crow. It has a more uniform iridescence above, and is greenish underneath. Its caw is hoarser and more nasal. Its range is from Connecticut and the lower Hudson southward, generally near the coast. It is abundant in Virginia, and near the city of Washington.

The Florida crow is similar to the American Crow, except that its bill and feet are larger, its wings and tail shorter.

The Raven

The northern raven so resembles the crow that it is often difficult to distinguish them. The chief differences are the raven's much greater size (from 22 to 26½ inches), and its note, which sounds more like *Croak* than *Caw*. This is the raven found in Alaska, northern Canada, and Greenland,—the bird especially revered by Alaskan Indians. It is found

[9] From "Birds of a Maryland Farm," by Sylvester D. Judd—Bulletin No. 17, U. S. Department of Agriculture, Division of Biological Survey.

also in the northern United States,—in the state of Washington, in Minnesota, the Adirondacks, and elsewhere.

Major Charles Bendire, in his "Life Histories of North 18 American Birds," makes the following statements about the northern raven:

> "It lives to a great extent on offal and refuse of any kind, and is generally most abundant in the immediate vicinity of Indian camps and settlements, which are mostly located on the seashore, or on the banks of the larger rivers in the interior where these birds act as scavengers. Hundreds of ravens may frequently be seen in the vicinity of the salmon-canning stations. Clams also form a part of their food; these are said to be carried some distance in the air and dropped on the rocks to break their shells. They also prey to no small extent on the young and eggs of different water-fowl."

THE CARDINAL

Cardinal Grosbeak, Redbird, Virginia Nightingale

(Cardinals belong to the Grosbeak group of the large Finch or Sparrow Family, or the *Fringillidæ*.)

Length: About 8¼ inches; slightly smaller than the robin.

General Appearance: Brilliant rose-red plumage; crested head and thick beak.

Male: A soft cardinal red, except for a black throat, a black band encircling bill, and, in winter, a grayish tinge to wings. Bill large, heavy, and light red. Red crest conspicuous; it may be raised and lowered at will. Tail long and slender; it is twitched nervously and frequently.

Female: Brownish-gray above, yellowish underneath. Crest, wings, and tail reddish—the color especially noticeable in flight. Throat and band about bill grayish-black.

Call-note: A sharp, insistent *tsip, tsip*.

Song: A loud and clear, yet sweet and mellow whistle, *cheer, cheer, he-u, he-u, he-u*, repeatedly rapidly with descending inflection, and with nearly an octave in range. The female, unlike most of her sex in the bird-world, is also a fine singer; her soft melodious warble is considered by many listeners to be superior to the song of her mate.

Figure 2. Cardinal.

Habitat: "Shrubbery is its chosen haunt, the more tangled the better. Here the nest is built and here they spend most of their days. Higher trees are usually sought only under the inspiration of song."[10]

[10] From W. L. McAtee; *Farmers' Bulletin 755*, Biological Survey, U. S. Department of Agriculture.

Range: From southeastern South Dakota, Iowa, northern Indiana and Ohio, southeastern and southwestern Pennsylvania, southern Hudson Valley, south to the Gulf States; a resident of Bermuda. Cardinals are not migratory.

Cardinals are especially numerous in our Southern States. They abound in Florida and Bermuda, where their brilliant coloring contrasts wonderfully with the light sands and the coral limestone. A cardinal singing in a hibiscus bush, laden with gorgeous red blooms, makes a never-to-be-forgotten memory; while a sight of one in a blossoming Virginia dog-wood tree or against a northern snow-scene is equally memorable. These birds are great favorites in the South, rivaling the mockingbirds in the affections of many people. In the North, a glimpse of a cardinal marks a red-letter day; and bird-lovers whose kind hands spread bountiful tables for winter residents, count themselves highly favored to have a pair of cardinals for their guests. Aside from the joy which their beauty and their song bring, they possess great practical value.

Mr. W. L. McAtee, of the Biological Survey, writes that about one-fourth of the cardinal's food consists of destructive pests such as the worms which infest cotton plants, and numerous other caterpillars, besides grasshoppers, scale insects, beetles, and others. A large part of their food consists of the seeds of troublesome weeds and of wild fruits. "The bird has a record for feeding on many of the worst agricultural pests."[11] No sins are laid at his door. "Cardinals are usually seen in pairs, but in 21 winter they often collect in southern swamps and thickets, and flock to feeding-places near the haunts of man when food is scarce."[12]

They were formerly trapped for cage-birds. They were so highly esteemed that they were in great demand even in Europe, where they received the name of the "Virginia Nightingale." But trapping is now nearly abolished, and the wild, liberty-loving cardinal may roam as he will with the wife of his heart. Few birds are more ardent, jealous lovers, more

[11] *Farmers' Bulletin 755*, Biological Survey, U. S. Department of Agriculture.

[12] William Dutcher, Former President of the National Association of Audubon Societies; *Educational Leaflet* No. 18.

tenderly devoted husbands, or more anxious, solicitous fathers than these beautiful, sweet-voiced redbirds.[12]

The Pine Grosbeak

Finch Family—Fringillidæ

Length: A little over 9 inches; slightly larger than his cousin, the cardinal, and nearly an inch smaller than the robin.

General Appearance: A *red bird* with *brown and white wings, a brown tail, and a heavy beak.*

Male: A bright raspberry-red, deepest on the *head*, *breast*, *rump*, and *upper tail-coverts*; the rest of the body a slaty gray, lighter underneath, with a *soft red breast*; wings dark brown, edged with white, forming two broad wing-bars; tail forked; beak large and strong, with a small hook at the end.

Female: Slaty gray, with head, rump, and upper tail-coverts *olive-yellow* where the males are red; under parts washed with yellow: wings and tail brown; wings edged with white; two wing-bars.

Young: Similar to female.

Song: A loud, clear whistle, given while on the wing. In spring, a melodious nesting song.

Range: Eastern North America. Breeds in the tree-regions of Canada, in the White Mts., and Maine; winters south to Iowa, Indiana, Pennsylvania, northern New Jersey, (and occasionally to the District of Columbia and Kentucky), westward to Manitoba, Minnesota, and Kansas.

This brilliant, handsome Pine Grosbeak is comparatively unknown in the United States, but wherever he appears as a rare visitor, he is hailed with enthusiasm or excitement because of his beautiful color. He resembles
23 his cousin, the purple finch, in color and markings, but is much larger.

Thoreau says, "When some rare northern bird like the pine grosbeak is seen thus far south in the winter, he does not suggest poverty, but dazzles us with his beauty. There is in them a warmth akin to the warmth that melts the icicle. Think of these brilliant, warm-colored, and richly-warbling birds, birds of paradise, dainty-footed, downy-clad, in the midst of a New England, a Canadian winter."[13]

The Pine Grosbeak

> "is of gentle, unobtrusive manner, almost entirely fearless of man's approach, and always seems to be perfectly contented with its situation wherever encountered. A whole tree full of these birds may be seen feeding on the seeds of mountain ash berries, apples, or the buds of beeches. One may stand within a few feet of them for a long time without their taking any notice of one's presence. They are slow and deliberate in manner. Their flight, however, is rather rapid and aggressive, slightly undulating."[14]

They are silent, uninteresting birds, awkward in their movements. They are very hardy, and roam southward when the severe Canadian winters send them forth in search of food. Seeds of cone-bearing trees, sumac and mountain ash berries are their favorite winter diet. They return to their northern nesting places when few birds would consider it seemly to set up housekeeping.

The American or Red Crossbill

Finch Family—Fringillidæ

Length: A little over 6 inches; slightly larger than the English sparrow.

General Appearance: A small, plump red bird, with brown wings, brown forked tail, and a *bill crossed at the tip*.

Male: Head and body a dull red, brownish on the back, and bright red above tail; wings brown, *without white bars*; tail brown and notched; bill with long strong mandibles that are crossed somewhat like a parrot's.

---

[13] From "Notes on New England Birds," H. D. Thoreau, page 421.
[14] From Eaton's "Birds of New York," page 255.

Female: Head and body dull olive, with a yellowish wash—brightest on rump; head, back, and under parts mottled with black.

Call-note: A short, clear, metallic whistle.

Song: A gentle warble, varied, and agreeable to hear.

Flight: Undulating.

Habitat: Coniferous forests, preferably.

Range: Northern North America. Breeds from central Alaska, and northern Canada south to the mountains of California, to Colorado, Michigan, and in the Alleghanies of Georgia, occasionally in Massachusetts, Maryland, and Virginia.

Red Crossbills are truly the "Wandering Jews" of the bird-world. They are erratic nomads, living in flocks, and roaming where fancy leads or necessity impels them. They pitch their tents and raise their broods wherever they may happen to be sojourning in late winter or early spring, even though many miles south of their natural breeding places. Dr. Elliot Coues writes: "Their most remarkable habit is that of breeding in the winter, or very early in the spring, when one would think it impossible that their callow young could endure the rigors of the season." He mentions a nest taken in Maine in February, and another in Vermont so early in March that the ground was covered with snow and the weather was very severe.[15]

They make no regular migrations, spring or fall, but like will-o'-the-wisps appear and vanish, affording one of the most delightful surprises to be found in nature. To see one of them, accompanied by his olive-green mate, swinging from a spruce bough against a flaming sunset sky or a snowy landscape, is an event in one's life.

Crossbills are denizens of coniferous forests. Their twisted or crossed bills are peculiarly adapted to extracting seeds from pine and spruce cones, though they eat berries, fruit, grass seeds, and cankerworms in season. Because of their curiously twisted beaks, these birds have always been regarded with peculiar interest, even with superstition. Longfellow has preserved for us the German legend regarding this bird in his poem:

---

[15] Educational Leaflet No. 35, National Association of Audubon Societies.

Figure 3. Crossbill.

The Legend of the Crossbill

On the cross the dying Saviour
Heavenward lifts his eyelids calm,
Feels, but scarcely feels, a trembling
In his pierced and bleeding palm.
And by all the world forsaken,
Sees He how with zealous care

At the ruthless nail of iron
A little bird is striving there.
Stained with blood and never tiring
With its beak it doth not cease;
From the cross 'twould free the Saviour,
Its Creator's Son release.
And the Saviour speaks in mildness:
"Blest be thou of all the good!
Bear, as token of this moment,
Marks of blood and holy rood!"
And that bird is called the crossbill;
Covered all with blood so clear,
In the groves of pine it singeth
Songs, like legends, strange to hear.[16]

Henry W. Longfellow

THE WHITE-WINGED CROSSBILL

The White-winged Crossbill is similar to the Red Crossbill, but its body is a *dull crimson instead of red*, and its black wing-feathers are so tipped with white as to form *two broad white wing-bars*. The female is olive-green, gray underneath, with a yellow rump, dark wings and tail, white wing-bars, and dark streaks on head, breast, and back.

This crossbill breeds in Canada, south to the Adirondacks, White Mountains, and Maine. Its note is a soft *cheep*; its song a gentle warble. To see a flock of these birds feeding silently in a grove of spruces or hear them singing their low sweet song makes a memory cherished by bird-lovers. They may be seen in winter as far south as North Carolina.

[16] Used with permission of the Houghton Mifflin Co., the authorized publishers.

THE JUNCO OR SLATE-COLORED SNOWBIRD

Finch Family—Fringillidæ

Length: About 6¼ inches; slightly smaller than the English sparrow.

General Appearance: Trim, dainty little birds, all gray and white, except for a *pinkish* or *flesh-colored* bill. *White outer tail-feathers*, showing in flight, are distinguishing marks.

Male: Dark slate-gray above and white below. The gray extends to the center of the breast in a nearly horizontal line, and with the white under parts, gives the effect of the birds' having waded breast-deep in the snow, or having been sliced in two, like the "sliced animals" of our childhood. Sides grayish; wings slightly darker; tail dark brown, with two outer feathers white; third feather, partly white; bill heavy, adapted to a diet of seeds.

Female: Similar to male, only brownish-gray. Winter plumage of all juncos browner than summer plumage.

Young: Light brownish, streaked with black.

Note: A gentle *tseep, tseep*, and a *smack, smack*, of alarm or distress.

Song: A tender, sweet trill in the spring. Though monotonous, the song is very pleasing.

Habitat: Groves of conifers; thickets of bushes or vines, or clumps of weeds.

Nest: Juncos' nests are built of mosses or grasses on or near the ground. The speckled eggs and the streaked babies are excellent examples of protective coloring. The nests are sometimes placed very near houses, if the surroundings are to the liking of the birds.

Range: Eastern and northern North America. Breeds from 28 the tree-limit of Alaska and Canada southward to northern United States,—northern Minnesota, central Michigan, Maine, the mountains of New York, Pennsylvania, and Massachusetts; winters throughout eastern United States and southern Canada to the Gulf Coast.

The Carolina Junco, nesting in the southern Appalachian mountains, is a subspecies, differing but very slightly in color.

Figure 4. Junco.

Juncos are gentle, attractive little creatures that come to our thickets when the chill of autumn has driven away our insectivorous birds. Being seed-eaters, they do not fear winter snows, except those that cover tall weeds. According to Professor Beal, juncos should be rigidly protected. They not only destroy large quantities of weed seeds, thereby rendering service to agriculture, but they eat harmful insects, of which caterpillars are their favorite. They do no damage to fruit or grain.[17]

Mr. Forbush writes of the junco as follows:

[17] Farmers' Bulletin 506, Biological Survey, U. S. Department of Agriculture.

"The Snowbird does not often breed in Massachusetts, excepting on the higher lands of the north-central and western parts of the State. Pairs are said to nest occasionally in ice-houses, which are certainly cool, if not suitable situations. It is a bird of the Canadian fauna, and it winters in Massachusetts whenever conditions are favorable. In the southeastern portion of the State, where the ground is bare in sheltered places through much of the winter, or where weed seed, chaff, and other food can be secured, this bird is common in the colder months. Its notes at this season are chiefly sparrow like chirps.

"A flock of these dark birds on the new-fallen snow is an interesting sight on a cold winter's day, as they come familiarly about the house or barnyard. Audubon says that in winter they burrow in stacks of corn or hay for shelter at night during the continuance of inclement weather. As spring comes they begin to sing much like the Chipping Sparrow. They converse together with a musical twittering, and about the first of May they leave for their northern breeding-ground."[18]

The Snowflake or Snow Bunting

Finch Family—Fringillidæ

Length: A little less than 7 inches; slightly larger than the junco and the English sparrow.

General Appearance: A brown, black, and white bird; the *white is conspicuous on wings and tail*, especially in flight. The bird has a characteristic way of "hugging the ground" when walking or running—it does not hop.

Male and Female: *In winter*: head brown on top, lighter on neck; white on sides of head, with a brown thumb-mark below eye; back brown, streaked with black; throat and belly white; a broad brownish band across breast; a brownish wash on sides and rump; wings black and white, some of the feathers edged with brown—in flight, the wings appear white,

[18] "Useful Birds and Their Protection," by E. H. Forbush.

broadly tipped with black; inner tail-feathers black, outer feathers white. *In summer*: back and shoulders black, the rest of the body white; wings and tail black and white.

Notes: Thoreau calls their note "a rippling whistle." He says also, "Besides their rippling note, they have a vibratory twitter, and from the loiterers you hear quite a tender peep."[19]

Habitat: The tundras of North America. Snow buntings breed in the treeless regions of the North; they migrate southward during the winter.

Range: Northern Hemisphere. In North America, they breed from 83° north (including Greenland), to the northern part of Canada and Alaska; winter from Unalaska and south-central Canada to northern United States, irregularly to northern California, Colorado, Kansas, southern Indiana and Ohio, and Florida.

Snow Buntings, or "Brown Snowbirds" as they are called to distinguish them from the juncos, or "Gray Snowbirds," are not generally known because of the infrequency and irregularity of their visits. They belong to the Sparrow family, but have so much black and white on their wings and tail as to appear very unlike their relatives.

Snowflakes are gentle, fearless little birds, possibly because they come from the sparsely settled regions of the North, where they need not learn to fear human beings. Like chickadees, they appear to love driving storms, and to frolic during February blizzards with as keen delight as warmly clad children; like tree sparrows, they are protected by a layer of fat that keeps out the cold. As they, too, are seed-eaters, snow buntings must journey southward during the winter to regions where deep snows do not bury the weeds.

Few people are aware that in the treeless plains of the north there lives a bird that resembles the much-admired skylark of England in its way of singing. Both snow buntings and skylarks begin to sing as they rise from the ground, sing while on the wing or high up in the air, then drop swiftly to the ground.

---

[19] From "Notes on New England Birds," H. D. Thoreau, page 278.

Figure 5. Snowflake.

Dr. Judd writes as follows about the snowbird:

“The snowflake is a bird of the arctic tundra, above the limit of tree growth. In North America it breeds about Hudson Bay, in the northernmost parts of Labrador and Alaska, and to the northward. In its northern home it is a white, black-blotched sparrow, of whose habits very little is 32 known, except that it makes a feather-lined nest on the ground, in which it rears four or five young on a diet which probably consists principally of insects. After the breeding season, however, a buffy brown comes mixed with the black and white, and the birds assume a more sparrowlike aspect. They migrate southward with the first severe cold weather, some of them coming as far south as the northern half of the United States, where their appearance is regarded as a sure sign that winter has begun in earnest. Often a flock of a thousand will come with a blizzard, the thermometer registering 30° to 40° below zero; and in their circling, swirling flight, as they are borne along by the blast, they might well be mistaken at a distance for veritable snowflakes. They settle in the open fields and along railroad tracks, where they secure some food from hayseed, grain that has sifted out of the grain cars, and seeds of weeds that grow along the tracks. Here they remain until April, when, in obedience to the migrating instinct, they journey north to nest on the treeless plains of the arctic regions.

"The snowflake differs from many other winter sparrows, such as the tree sparrow, junco, and white-throated sparrow, in that its flocks act more nearly as units, the alarm of a single member causing the whole flock to whirl up into the air and be off. A further difference may be noted in its strictly terrestrial habits. When not flying, it is almost invariably found on the ground; and when it does happen to alight in a tree, awkward wobblings betray its discomfort. Where the feeding conditions are favorable, immense flocks of snowflakes may be seen apparently rolling like a cloud across the land, this curious effect being 33 due to the rear rank continually rising and flying forward to a point just in advance of the rest of the flock."[20]

Dr. Judd says that little information can be given concerning the summer food of this bird, but that it probably feeds on the seeds of shore or marsh plants. The winter food consists of grain, mostly gleanings or waste, and of weed seed which is consumed in enormous quantities. "On account of its good work as a weed destroyer and the apparent absence of any

---

[20] From "The Relation of Sparrows to Agriculture," by Sylvester D. Judd, Bulletin No. 15, Biological Survey.

noticeably detrimental food habits, the snowflake seems to deserve high commendation, and should receive careful protection."

THE TREE SPARROW OR WINTER CHIPPY

Finch Family—Fringillidæ

Length: A little over 6 inches; about the size of the English sparrow.

General Appearance: A small brown bird with a gray breast that has an *indistinct black spot* in the *center*.

Male and Female: *Crown reddish-brown*; a gray line over the eye, a reddish-brown line back of eye; gray below eye; a reddish-brown streak curving from bill; bill short and thick; back brown, streaked with black and buff; wings dark brown, edged with white, and with two white wing-bars; tail brown, slightly forked, outer feathers edged with white; sides brownish, other under parts white; *the black spot in the center of the breast*, the identification mark.

Notes: Cheerful twitters and chirps.

Song: A sweet, gentle trill, very delightful to hear.

Habitat: Fields, especially those bordered by bushes that can be used as shelter at night and as a refuge from enemies.

Range: Eastern North America. Breeds in northern and central Canada; winters from southern Minnesota and southeastern Canada to eastern Oklahoma, central Arkansas, and South Carolina.

The Tree Sparrow

When lordly Winter stalks abroad
With trailing robes of snow,
That hide the lovely tender things
His icy breath lays low;
When grasses, shrubs, and hardy weeds

Hold high their heads, and mock
Their tyrant lord,—from Northland woods
There come a merry flock
Of feathered songsters, soft and brown
With a dark spot on each breast.
They sway on stalk of golden-rod
Above a snowdrift's crest.
Their voices ring like tinkling bells
Beneath the wintry sky,
Till April, when with joyous songs
Back to the North they fly.

Such are the rollicking little Tree Sparrows, that whirl into our vision like an eddy of brown leaves. To the untrained observer, they are "just sparrows," but to the "seeing eye" they are altogether more dainty and refined than English sparrows, and have different markings. Their little brown caps, the gray line over their bright eyes, their brown backs, white wing-bars, pale gray breasts and forked tails resemble those of their little cousins, the chipping sparrows. But *the soft grayish-black spot on each* tree sparrow's breast is a difference. Careful comparison with the "Chippy" will show no straight black line extending from the eye, but a brown curve behind the eye that joins the one extending from the bill.

The voices of winter chippies are infinitely sweeter than those of the door yard chippies and their English relatives. Their note is sweet and joyous. Mr. Forbush writes of their song as follows:

> "Tree Sparrows are among the few birds that can 'look our winters in the face and sing.' They are occasionally heard singing in November and December and late in February, when deep snow covers the ground. The song is among the sweetest of sparrow notes, but not very strong. It slightly resembles that of the Fox 36 Sparrow. Like other sparrows they chirp and twitter from time to time, but the full chorus of a flock in winter is a sound worth going far to hear."[21]

---

[21] From "Useful Birds and Their Protection," by E. H. Forbush.

Figure 6. Tree sparrow.

Dr. Judd says:

"The tree sparrow breeds in Labrador and the Hudson Bay region and westward to Alaska. In the fall the birds come down from the north in immense throngs and spread over the United States as far south as South Carolina, Kansas, and Arizona. During the winter, in company with juncos, white-throats, white-crowns, and fox sparrows, they give life to the hedgerows, tangled thickets, and weed patches.... The food of the tree sparrow during its stay in the United States is almost entirely made up of seeds. The bird shows an essential difference from its associates,

> however, in its large consumption of grass seed, fully half of its food consisting of this element.... Nearly two-thirds of the vegetable food that is not grass seed is derived from such plants as ragweed, amaranth, lamb's quarters, ... and a variety of seeds such as wild sunflower, goldenrod, chickweed, purslane, wood sorrel, violet, and sheep sorrel."[22]

Professor Beal says that the oily seeds of such plants as ragweed cause the little bodies of tree sparrows to be encased in "a layer of fat constituting a set of under-flannels from one-eighth to one-fourth of an inch in thickness all over the bird's body." They are so warmly dressed that it is no wonder they are happy, cheerful, and active. A sight of them in a beautiful, snowy meadow is enough to repay one for the trouble of a quest.

Pine siskins, redpolls, song sparrows, white-throated sparrows, purple finches, and goldfinches are other species of the large Finch family, or Fringillidæ, that may be seen during the winter months.

The Pine siskin or pine finch is a small brownish-gray bird streaked with black, and with *buff edges* to many of its feathers. The *yellow in the wings and in the forked tail* will distinguish it.

The redpoll is a little brownish-gray bird with a red forehead, reddish breast and rump, black chin and throat. It has distinct dark streaks on its head, back, and under parts, except the breast. There are several species varying slightly in size and markings.

The Bobwhite or Quail

American Partridge Family—Odontophoridæ

Length: About 10 inches; the same length as the robin, but the quail has a stouter body and a shorter tail.

General Appearance: A plump, mottled brown bird, with a small head, short bill, and short tail.

---

[22] Bulletin No. 15, U. S. Dept. of Agriculture, Biological Survey.

Male: Upper parts reddish-brown and chestnut-brown, mottled with black, gray, and buff; head slightly crested; *forehead and line above eye white*, line extending to neck; black patch below eye, that curves to enclose *white throat* and forms a band below it; under parts whitish, barred with black, except upper part of breast which is reddish-brown; tail short, gray, mottled with buff and a few black flecks.

Female: Similar to male, except for *buff patch over eye and buff throat*, and less black on head, neck, and across breast. In summer, the crown of both sexes is darker than in winter; the buff markings are lighter in color.

Note: *Bob-white? Bob-bob-white?* A very clear, sweet, musical whistle.

Habitat: Grassy meadows and cultivated fields; farmyards, thickets, and swamps during the winter.

Range: Eastern North America, from southern Canada to the Gulf Coast and northern Florida and west to eastern Colorado. Usually a resident.

In Florida, except in the north, is found the Florida bobwhite, a smaller and darker species. A quail is called a partridge in the south. The California quail, one of several western species, is very different in appearance from the eastern quail. It has a nodding plume on its head and is largely black, white, and brownish-gray.

No birds of my acquaintance, unless it be bluebirds, goldfinches, chickadees, and thrushes, seem so lovable, so interesting, and so altogether desirable as quail. Our summer meadows would lose much of their charm without the cheery "Bob White" ringing across them.

The character of human beings is shown in their voices; that of birds seems likewise revealed. The note of the quail breathes sweetness, tenderness, joy in life, and deep contentment. Unless need of food compels it, the killing of these nearly human creatures seems to me like the "Slaughter of the Innocents."

Few birds are so devoted to their mates or to their young as the quail. Many human parents are less alive to parental responsibilities. It is a well-known fact that while Mother Quail is sitting upon her second nestful of a dozen or more eggs, Father Bob assumes the entire care of the large, restless, older brood.

Figure 7. Bob-white.

Most birds love their mates and their young, but quail seem to have affection for their brothers and sisters, also. The parents and the two broods sometimes remain together during the winter. When one member of the family is lost, the others give their tender covey-call, to lure home the prodigal. There are few sweeter sounds in nature. Mr. Forbush says:

> "When the broods are scattered by the gunner, they are reassembled again by a whistled call of the old bird, which has been given, '*ka-loi-kee, ka-loi-kee*,' and is answered by the whistled repeated response, '*whoil-kee*.' The syllables almost run together. The first call is uttered with a rising and the other with a falling 40 inflection. It is plainly the rallying call and the answering cry."[23]

Dallas Lore Sharp, in his charming book "Wild Life Near Home," refers to the covey-call as follows:

> "It was the sweetest bird-note I ever heard, being so low, so liquid, so mellow that I almost doubted if Bob White could make it. But there she stood in the snow with head high, listening anxiously. Again she whistled, louder this time; and from the woods below came a faint answering call, *White!* The answer seemed to break a spell; and on three sides of me sounded other calls. At this the little signaler repeated her efforts, and each time the answers came louder and nearer. Presently something dark hurried by me over the snow and joined the quail I was watching. It was one of the covey I had heard call from the woods.
>
> "Again and again the signal was sent forth, until a third, fourth, and finally a fifth were grouped about the leader. There was just an audible twitter of welcome and gratitude exchanged as each new-comer made his appearance. Once more the whistle sounded; but this time there was no response across the silent field."

Young quail are very precocious. They are able to run about soon after they are hatched. They early learn how to hide and "freeze." A friend told me of coming suddenly upon a brood. The mother gave a call and all fled instantly, except one that turned into a little brown wooden image under a leaf at his feet. He picked it up and held it in his hand. Not a motion did it make until its mother gave a second call, when it shot out of his hand like a flash.

Another friend told me of her experience in finding a lost baby-quail. It was too little and too weak to keep up with the family—was probably the

---

[23] "Useful Birds and Their Protection," E. H. Forbush, page 328.

last born. It was so tired and distressed that when she knelt down and placed her cupped hand near it, the poor little thing ran to it, nestled down, and shut its eyes. She discovered the brood and carried the baby over to join its family, but it seemed loath to leave her. Three times it ran back to the warm shelter of her hand. She could hardly bear to abandon it to the life that seemed more than it could endure.

Dr. Judd made a careful study of the bobwhite. The following extracts are from his report:

> "It is the general opinion that with the on-coming of winter the bobwhite is found less often in the open fields, when withered herbaceous plants afford but scant protection from enemies, than in dense bushy, briery coverts and woods. In Maryland and Virginia, the scattered and depleted coveys after the shooting season evidently unite into large bevies. Their favorite resort is a bank with a southern exposure and suitable food-supply.
>
> "Robert Ridgway found a clutch of freshly deposited eggs in southern Illinois on October 16, and H. C. Munger found another set in Missouri in January, the parent being afterwards found frozen on the nest. Authentic records show that bobwhite has been known to breed, at least occasionally, somewhere in its range every month in the year....
>
> "In Maryland and Virginia large land-owners often feed their birds in severe weather. Wheat and corn are the best food and should be scattered, if possible, among the briers where the birds are safe from hawks. Bobwhites 42 have been known to feed with chickens in barnyards. By a little forethought land-owners and sportsmen can easily make provision for their birds. Sumac bushes should be left along hedgerows and the edge of woodland to furnish food that is always above the snow and lasts well into spring.... The bayberry and wax-myrtle last until May, also.
>
> "The food habits of the bobwhite are noteworthy in several respects. Vegetable matter has long been known to be an important element in the food of the bobwhite. Grain-eating birds are likely to do much harm to crops.... The bobwhite is a notable exception. Not a single sprouting kernel was found in the crops and stomachs of quail examined."[24]

---

[24] Bulletin No. 21, Bureau of Biological Survey, U.S. Department of Agriculture.

Dr. Judd enumerates eighty-eight varieties of weed seeds that are eaten by quail, and states an amazing number eaten at one time.

> "One bird shot at Marshall Hall had eaten 1000 ragweed akenes; another contained [quantities of] leguminous seeds, mainly tick-trefoil; a third had eaten 5000 seeds of green foxtail grass, while a fourth had taken about 10000 [infinitesimal] pigweed seeds."[24]

As an insect-destroyer the bobwhite is of enormous value. During the summer, insects form more then one-third of its food. Over one hundred varieties had been discovered by examination of the stomachs of quail in 1905, an unusually large proportion of which were highly injurious to crops. Mr. Forbush thinks that no farmer in Massachusetts can afford to shoot a quail or allow it to be shot on his land, and that if the markets must be supplied, quail must be reared artificially.

Our bobwhite sleeps on the ground. The California quail roosts in bushes or trees. One summer evening in Santa Barbara it was my privilege to see a charming phase of quail family life. I was sitting quietly under a tree on a knoll that overlooked a flat shed-roof, when I heard a low call, and a whirring of wings. Mother Quail, accompanied by thirteen little balls of brown feathers, alighted on the roof near me. She talked to her adorable family, and, judging by their quick responses, she evidently gave them numerous commands. They finally ran to the edge of the roof and arranged themselves in a row, faces outward, until she gave another call. Then obediently they gathered around her in a true Kindergarten Circle, heads outward and tails toward her, all ready for bed. There they nestled, until a passer-by disturbed them and, to my great regret, they flew away. In a few minutes I heard a clear loud *ku-ku-kow*, and on the same roof alighted Father Bob with fifteen restless boys and girls—a veritable Primary Class. He had more trouble in controlling them than Mother had experienced with her docile little ones; they ran hither and thither in spite of his insistent, anxious calls. He succeeded in gathering them about him, however; but just as they were forming their circle, they, too, were frightened away.

The Ruffed Grouse

Grouse Family—Tetraonidæ

Length: About 17 inches.

Male: Upper parts reddish-brown, with black, yellowish, gray, and whitish markings; large tufts or "ruffs" of glossy black feathers at the sides of the neck. Tail long and broad, gray and reddish-brown, mottled and barred with black, and a broad blackish band near the end; when spread, the tail resembles a fan. Under parts buffy, becoming white, with black bars that are indistinct on breast and belly, and darker on the sides; a broken band on the breast.

Female: Similar to male, but with smaller ruffs on the sides of the neck.

"Love-song": A loud tattoo or drumming that sounds like a thump on a large drum—a *tum-tum-tum-tum-tum-tup-tup-whir-r-r-r-r-r*. This tattoo is most common in late winter and early spring, but may be heard in the summer and fall. While heard most frequently during the day, it may be heard at any hour of the night. In making it, the bird usually stands very erect on a hollow log or stump, with head held high and ruffs erected and spread, and, raising its wings, strikes downward and forward. The sound produced is a muffled boom or thump. It begins with a few slow beats, gradually growing quicker, and ends in a rolling, accelerated "tattoo."[25]

Habitat: A bird of the woods that nests on the ground.

Range: A resident in the northern two-thirds of the United States and in the forested parts of Canada.

The Ruffed Grouse, the finest and most famous game-bird of the northern woods, was formerly very abundant. Its numbers have greatly decreased. Like the bobwhite, it responds to protection and may be raised under artificial conditions. It eats nearly sixty kinds of wild fruit; beechnuts, hazelnuts, chestnuts, acorns, and weed seeds form a large part

[25] From "Useful Birds and Their Protection," by E. H. Forbush.

of its diet. It eats some insects, the most important being beetles of various kinds.

Mr. Forbush says:

> "The female alone undertakes the task of incubation and the care of the young.... All the young grouse in a nest hatch at nearly the same instant; their feathers dry very rapidly, and they are soon ready to run about.... They run about, stealing noiselessly along among the dead leaves, under the foliage of ferns and shrubbery.... Meanwhile, the mother marches slowly in the rear, perhaps to guard them against surprise from any keen-scented animal that may follow on the trail. She seems to be always on the alert, and a single warning note from her will cause the young birds to flatten themselves on the ground or to hide under leaves, where they will remain motionless until they are trodden upon, rather than run the risk of betraying themselves by attempting to escape.
>
> "During the fall, the Grouse keep together in small flocks. Sometimes a dozen birds may be found around some favorite grape vine or apple tree, but they are usually so harried and scattered by gunners that toward winter the old birds may sometimes be found alone.
>
> "As winter approaches, this hardy bird puts on its 'snowshoes,' which consist of a fringe of horny processes or pectinations that grow out along each toe, and help to 46 distribute the weight of the bird over a larger surface, and so allow it to walk over snows into which a bird not so provided would sink deeply. Its digestion must resemble that of the famous Ostrich, as broken twigs and dry leaves are ground up in its mill. It is a hard winter that will starve the Grouse. A pair spent many winter nights in a little cave in the rocky wall of an old quarry. Sumacs grew there, and many rank weeds. The birds lived well on sumac berries, weed seeds, and buds.
>
> "Sometimes, but perhaps rarely, these birds are imprisoned under the snow by the icy crust which forms in cold weather following a rain, but usually they are vigorous enough to find a way out somewhere. The Grouse is perfectly at home beneath the snow; it will dive into it to escape a Hawk, and can move rapidly about beneath the surface and burst out again in rapid flight at some unexpected place.
>
> "The Ruffed Grouse is a bird of the woodland, and though useful in the woods, it sometimes does some injury in the orchard by removing too

many buds from a single tree. In winter and early spring, when other food is buried by the snow and hard to obtain, the Grouse lives largely on the buds and green twigs of trees; but as spring advances, insects form a considerable part of the food. The young feed very largely on insects, including many very destructive species."[26]

The Cedar Waxwing or Cedar-Bird

Called Locally the "Cherry Bird" Waxwing Family or Bombycillidæ

Length: A little over 7 inches.

General Appearance: A grayish-brown bird, with a decided crest and a yellow band at end of tail. Plump and well-fed in appearance.

Male and Female: A beautiful, rich grayish-brown with a soft yellow breast. Head conspicuously crested; forehead glossy black; a black line above the bill is extended toward the top of the head, outlining the crest; crest elevated and lowered to express surprise, contentment, fear and other emotions; bill and chin black; throat blackish. Wings brown, becoming a soft gray; wing-feathers with small red tips that look like bits of sealing-wax—hence the name, Waxwing. Tail light gray, shading to a dark grey, rounded, fan-shaped in flight, and edged with a broad yellow band.

Young: Grayish-brown, streaked, and without red tips to their wings.

Note: A gentle lisping *tseep, tseep*, monotonous and uninteresting. Mr. Forbush says of the waxwing, "It moves about in silence, save as it utters a lisping 'beading' note or a 'hushed whistle.'"

Habitat: During the nesting season, devoted pairs may be seen in orchards, in red cedars, or in shrubbery by roadsides, preferably near trees or bushes laden with berries. The birds are rovers, usually flying in large flocks.

[26] From "Useful Birds and Their Protection," by E. H. Forbush.

Range: North America. Breeds from south-central Canada to southern Oregon, northern New Mexico, Kansas, 48 northern Arkansas, and North Carolina; winters irregularly throughout nearly all the United States, and south to Cuba, Mexico, and Panama.

Figure 8. Cedar waxwing.

Cedar Waxwings are among our most exquisite birds in their delicate blending of color and in their dainty refinement. They seem to have been tinted by a water-color artist, or an expert in the use of pastels. Their proverbial good manners seem to preclude any disturbance of their well-preened feathers by undue haste of movement or quarrelsome ruffling.

My earliest recollections of these beautiful but rather uninteresting birds is of their frequent raids upon a great mulberry tree in my grandparents' garden. They gorged upon the dead-ripe mulberries with the quiet enjoyment of epicures rather than the greedy haste of gourmands. I remember, also, my grandmother's dismay at the inroads which the "cherry-birds" and robins made upon her cherry crop, and my bird-loving grandfather's command that no bird should be molested.

Cedar, juniper, sumac, and mountain ash berries, form the winter diet of these frugivorous birds. As a larder is speedily exhausted by a flock of from twenty to sixty hungry fruit-eaters, they must fly to "pastures new." During the spring and summer seasons, they supplement their diet of wild fruit, most of which is of no commercial value, with beetles that infest potato-patches and elm trees, and cankerworms that prey upon apple trees. They are very valuable to man, and earn their dessert of cultivated cherries. Mr. Forbush says that they deserve the name of "cankerworm birds."

He writes as follows:

> "They frequent infested orchards 49 in large flocks, and fill themselves with the worms until they can eat no more. Such little gluttons rarely can be found among birds. The Cedar-bird seems to have the most rapid digestion of any bird with which experiments have been made. Audubon said that Cedar-birds would gorge themselves with fruit until they could be taken by hand; and that he had seen wounded birds, confined in a cage, eat of apples until suffocated. They will stuff themselves to the very throat. So, wherever they feed, their appetites produce a visible effect. Professor Forbes estimates that thirty Cedar-birds will destroy ninety thousand cankerworms in a month. This calculation seems to be far within bounds.
>
> "Cedar-birds are devoted to each other and to their young. Sometimes a row of six or eight may be seen, sitting close together on a

limb, passing and repassing from beak to beak a fat caterpillar or juicy cherry. I have seen this touching courtesy but once, and believe it was done not so much from politeness as from the fact that most of the birds were so full that they had no room for more—a condition in which they could afford to be generous. Nevertheless, the manner in which it was done, and the simulation of tender regard and consideration for each other exhibited, rendered it a sight well worth seeing. They also have a habit of 'billing' or saluting one another with the bill."[27]

A flock of cedar-birds "seep" and whisper to each other like over-fed children. Their note seems to be an expression of their gentle, affectionate, comfortable, ease-loving natures. There appears to be absence of aspiration or 50 longing in their bird-hearts, which seems so poignant in thrushes and many other songsters.

The Bohemian Waxwing

The Bohemian Waxwing is very similar to its cousin, the Cedar Waxwing, in color and markings, but may be distinguished by its *larger size*, (8 inches), by *reddish-brown feathers under the tail, by the absence of yellow on the breast, by a crown that is reddish-brown in front*, and *by yellow and white markings on the wings*. In note, feeding habits, and other characteristics, it resembles the Cedar-bird.

This larger species of waxwing is found in the colder regions of the whole Northern Hemisphere. In North America it breeds from northern Alaska and northern Canada to southern British Columbia and Alberta; winters east to Nova Scotia and south irregularly to eastern California, Colorado, Kansas, southern Illinois, Indiana, Ohio, Pennsylvania, and Connecticut. It is a rare winter visitor in Massachusetts.

[27] "Useful Birds and Their Protection," by E. H. Forbush, page 210.

THE TUFTED TITMOUSE OR TOMTIT

Titmouse Family—Paridæ

Length: About 6 inches; a little smaller than the English sparrow.

General Appearance: A slender, active, gray and white bird, *with a crest*. Its reddish-brown sides are not visible at a distance. The titmouse need never be confused with the waxwing; it is much smaller, and lacks the yellow and red markings on tail and wings.

Male and Female: Head conspicuously crested; crest gray and pointed; forehead black; bill short, sharp, black; back, wings, and tail gray; under parts whitish, with a reddish-brown wash on the sides.

Call-note: *De-de-de-de*, similar to one of the chickadee's notes, but louder.

Song: A loud, sweet, clear whistle: *Pe'-to, pe'-to, pe'-to, pe'-to, pe'-to*, frequently repeated five times. The titmouse is called locally the "Peter-bird."

Habitat: Woodlands; open groves of hard-wood trees preferred.

Range: Rare in New England. From Nebraska, Iowa, Illinois, Indiana, Ohio, Pennsylvania, and New Jersey, south to central Texas, the Gulf Coast, and Florida; occasional in Wisconsin, Michigan, New York, and Connecticut. Common permanent resident near Washington, especially in winter.

No winter bird more truly exemplifies protective coloring than the lively crested Tomtit, unless it be his little cousin, the Black-capped Chickadee. This sober-hued titmouse is such a blending of the grays and blacks of tree-trunk and icy brook, of the dazzling white of 52 snow and the soft gray shadows that lie across it, of reddish-brown shrubs and weeds, that he might escape notice except for his conspicuous crest. He can be distinguished from the cedar waxwing at a glance by his reddish sides, and because of the *absence* of a yellow band across the tail and of conspicuous black, white, and red patches or markings.

Figure 9. Tufted titmouse.

Few more active birds exist than titmice. They are at once the envy and the despair of aspiring small boys who know them, because of their extreme agility—their ability to perform acrobatic feats. They swing head downward from twigs in the search for their favorite food of insect-eggs; they seem strung on wires.

In the woodlands frequented by tufted titmice, they are as much in evidence as blue jays, because of their loud, clear *peto-peto-peto-peto-peto*, a welcome and pleasant sound during belated spring days or a bleak March "sugaring-off" season.

They are less friendly than chickadees, but are not shy, so they can be observed easily. They are very sociable with their kind, and are found, "playing around" with chickadees, nuthatches, and downy woodpeckers in the winter-time, and snuggling close together in old nest-holes during winter weather. In the spring, titmice use hollowed trees for their nesting sites and have been known to welcome a nesting-box.

These birds do enormous good, not only in eating insect-eggs, but in destroying caterpillars, cutworms, beetles, weevils, flies, wasps, plant-lice, and scale-insects in their season.[28] They will eat berries, nuts, and acorns during the winter and are extremely hardy.

The Chickadee or Black-Capped Titmouse

Titmouse Family—Paridæ

Length: About 5¼ inches.

General Appearance: A very active little gray and white bird, with a black cap and throat and dull yellowish sides.

Male and Female: Head and throat a glistening black; sides of head white; bill small, black, sharp-pointed; back a soft brownish-gray; wings and tail gray, edged with white; breast white, becoming yellowish at the sides below the wings.

Song: *Chick-a-dee-dee-dee*, uttered with gurgles and chuckles, and with variations.

Call-notes: *Day'-day*, and a whistle that resembles the word *Pé-whee*. The latter note is often called the "Phœbe note," and sometimes the "Pewee note." To me it resembles neither; it is not hoarse and wheezing like the phœbe's, nor plaintive like the pewee's. The last syllable has a descending inflection.

Flight: Very swift and jerky.

Habitat: Woodlands, orchards, and groves.

---

[28] *Farmers' Bulletin 630*, Biological Survey, U. S. Department of Agriculture.

Range: Eastern North America, from the Hudson Bay region and N. F., south to central Missouri, Illinois, northern Indiana, Ohio, Pennsylvania, northern New Jersey, and in the Alleghany Mts. to North Carolina; somewhat farther south in winter.

Figure 10. Chickadee.

The Carolina chickadee, a smaller species, breeds from central Missouri, Indiana, central Ohio, Pennsylvania (infrequently), and central New Jersey, south to southeastern Louisiana, the Gulf Coast and northern

Florida. In southern Florida, are found the Florida chickadees, that are still smaller and browner.

In the White Mts., the Green Mts., the Adirondacks, and southeastern Canada live Acadian chickadees, that differ from the preceding species in having brownish-gray crowns, and reddish-brown sides. A similarly marked species, slightly larger, is found from Ontario to Alaska.

During tiresome days of a winter convalescence, spent largely on a sleeping-porch that overlooked a beautiful hillside, my most constant and cheering companions were lively little chickadees. Their blending with the winter landscape was perfect. Whether they were seen against the black snow-laden trunks or smooth gray boles of beeches, or among yellowish willow-withes, they were bits of color harmony.

These active little gymnasts, performing unexpected feats in their swinging from horizontal bars, furnished pleasant diversion, while their friendly, confiding ways, their undaunted fearlessness, and their optimism cheered lonely hours.

An ice-storm necessitated the spreading of a table for our brave little all-kinds-of-weather friends. They came in pairs, grew very tame, and drew near to us like confiding children who knew that no harm would befall them. They acted as though our care of them was the most natural thing in the world. Chickadees have never seemed to me to "grow up," but always to remain the trusting little ones of the bird-world, too small to be out alone, and yet, like children, to fare forth with confidence that their needs would be supplied.

They repay a thousand-fold any care bestowed upon them. Dr. Judd reported finding in the stomach of one black-capped chickadee between 200 and 300 eggs of the 55 fall cankerworm moth, and 450 eggs of a plant louse in another. Mr. C. E. Bailey computed that one chickadee alone would destroy 138,750 eggs of the cankerworm moth in 25 days, while Prof. Sanderson estimated that 8,000,000,000 insects are destroyed yearly in Michigan by these invaluable little birds.[29]

---

[29] *Educational Leaflet* No. 61, National Association of Audubon Societies.

"Much of the daylight life of the chickadee is spent in a busy, active pursuit of, or search for, insects and their eggs. This is particularly the case in winter, when hibernating insects or their eggs must be most diligently sought, for then starvation always threatens. But the chickadee is one of the few insectivorous birds that is keen-witted enough to find abundant food and safe shelter during the inclement northern winter. Nevertheless, its busy search for food is sometimes interrupted for so long a time during severe storms, when the trees are encased in ice, that it dies from cold and hunger. During a sleet storm Mr. C. E. Bailey saw two chickadees creep under the loose clapboards of an old building for shelter. Their tails were so weighted down with ice that they could hardly fly, and had he not cared for them they might have perished.

"The chickadee, notwithstanding its hardiness, requires protection from cold winds and storms at night. It finds such shelter either in some hollow tree or in some deserted bird nest. Late one cold and snowy afternoon Mr. Bailey detected a movement in a cavity under an old crow's nest, and on climbing the tree he found two chickadees nestling there. They remained there until he had climbed to the nest and put his hand on one, when they flew out, only to return before he reached the ground. Minot speaks of a 56 chickadee that slept alone in winter in a phoebe's nest under his veranda. It retires to its refuge rather early at night, and does not come out until the Tree Sparrow, Song Sparrow, and Junco are abroad."[30]

THE GOLDEN-CROWNED KINGLET

Old World Warbler Family—Sylviidæ

Length: About 4 inches; smaller than the chickadee.

Male: Olive-green above, grayish-white underneath; crown with a *bright red center, bordered on each side* by *bright yellow*, and by a *black stripe* that edges the yellow; a light line over the eye; wings and tail brown; tail forked.

---

[30] "Useful Birds and Their Protection," E. H. Forbush, page 166.

Female: Like male, but without the red in the center of the yellow-and-black crown.

Call-note: A weak *tzee, tzee*, highly pitched.

Song: William Brewster, in the *Auk* for 1888, describes the song as follows: [It] "begins with a succession of five or six fine, shrill, high-pitched somewhat faltering notes, and ends with a short, rapid, rather explosive warble. The opening notes are given in a rising key, but the song falls rapidly at the end. The whole may be expressed as follows: *tzee, tzee, tzee, tzee, ti, ti, ter, ti-ti-ti-ti*."

Habitat: Woodlands, where kinglets are usually found *near the ends of branches*, of coniferous trees especially.

Range: Eastern North America. Breeds in the tree-regions of central Canada, south in the Rocky Mts. to northern Arizona, New Mexico, and to Michigan, New York, and mountains of Massachusetts, and in the higher Alleghanies south to North Carolina; winters from Iowa, Ontario, New Brunswick, to northern Florida and Mexico.

Though the Golden-crowned Kinglet is one of our smallest birds, it braves the rigors of winter in the United States. It may be seen from the latter part 58 of September until April or early May, when it goes to its more northerly nesting ground.

Kinglets and chickadees are industrious searchers for insects' eggs. Their value is almost inestimable. Mr. Forbush tells of watching the "Gold-crest" hunt for its food among the pines. He says:

> "The birds were fluttering about among the trees. Each one would hover for a moment before a tuft of pine 'needles,' and then either alight upon it and feed or pass on to another. I examined the 'needles' after the Kinglets had left them, and could find nothing on them; but when a bird was disturbed before it had finished feeding, the spray from which it had been driven was invariably found to be infested with numerous black specks, the eggs of plant lice. Evidently the birds were cleaning each spray thoroughly, as far as they went."[31]

---

[31] "Useful Birds and Their Protection," by E. H. Forbush, pages 161, 162, 163.

Mr. Forbush tells also of observing the work of seven kinglets in a grove of white pine which

> "must have been infested with countless thousands of these eggs, for the band of Kinglets remained there until March 25, almost three months later, apparently feeding most of the time on these eggs. When they had cleared the branches, the little birds fluttered about the trunks, hanging poised on busy wing, like Hummingbirds before a flower, meanwhile rapidly pecking the clinging eggs from the bark. In those three months they must have suppressed hosts of little tree pests, for I have never seen birds more industrious and assiduous in their attentions to the trees. One might expect such work of Creepers or of Woodpeckers; but the Kinglets 59 seemed to have departed from their usual habits of gleaning among limbs and foliage, to take the place of the missing Creepers, not one of which was seen in the grove last winter."[32]

The Carolina Wren

Wren Family—Troglodytidæ

Length: About 5½ inches; the largest of the six more common eastern wrens.

Male and Female: Reddish-brown above; *no bars* or *streaks*, except on wings and tail, and occasionally underneath the body, near the tail; a *long light line over the eye, extending to the shoulders*; under parts buff with a brownish wash; throat white.

Notes: "Wren-like *chucks* of annoyance or interrogation," and "a peculiar fluttering *k-r-r-r-r-uck*, which resembles the bleating call of a tree-toad."[33]

Song: A loud clear whistle, consisting of three similar syllables, with variations.

Habitat: Thickets, vines, and undergrowth.

---

[32] "Useful Birds and Their Protection," by E. H. Forbush, pages 161, 162, 163.

[33] From Witmer Stone in Educational Leaflet No. 50, National Association of Audubon Societies.

Range: Eastern United States. Breeds from southeastern Nebraska, Iowa, Ohio, southern Pennsylvania, the lower Hudson and Connecticut valleys south to central Texas, Gulf States, and northern Florida; casual north to Wisconsin, Michigan, Ontario, Massachusetts, New Hampshire, and Maine.

Professor Beal writes of this interesting wren as follows: "The Carolina wren is resident from the Gulf of Mexico north to the southern boundaries of Iowa, Illinois, and Connecticut in the breeding season, but in winter it withdraws somewhat farther south. It is a bird of the thicket and undergrowth, preferring to place its nest in holes and crannies, but when necessary, will build 61 a bulky structure in a tangle of twigs and vines. Unlike the house wren it does not ordinarily use the structures of man for nesting sites.

"It is one of the few American birds that sing throughout the year. Most birds sing, or try to, in the mating season, but the Carolina wren may be heard pouring forth his melody of song every month. The writer's first introduction to this bird was in the month of January when he heard gushing from a thicket a song which reminded him of June instead of midwinter.

"This wren keeps up the reputation of the family as an insect-eater, as over nine-tenths of its diet consists of insects and their allies." Stomach analysis shows that the vegetable food of the Carolina wren is largely seeds of trees and shrubs and some wild berries. He concludes: "From this analysis of the food of the Carolina wren, it is evident that the farmer and fruit-grower have not the slightest cause for complaint against the bird. It eats neither cultivated fruit nor grain, and does not even nest in an orchard tree; but it does feed on numerous injurious insects and enlivens the tangled thickets with its cheerful songs for twelve months of the year."[34]

Dr. Witmer Stone writes of the song of the Carolina wren as follows: "His most characteristic song has been likened by Mr. Chapman to *tea-kettle, tea-kettle, tea-kettle*, and to *whee-udle, whee-udle, whee-udle*. Wilson wrote it *sweet-william, sweet-william, sweet-william*; and to

[34] Farmers' Bulletin 755, U. S. Dept. of Agriculture, Biological Survey.

Audubon it seemed to say *come-to-me, come-to-me, come-to-me*. It has variations recalling forms in the Cardinal's song, and also that of the Tufted Titmouse; and the Wren after 62 repeating one form for some time, often changes suddenly to another producing a rather startling effect, as if another bird has taken its place."[35]

The Winter Wren

Wren Family—Troglodytidæ

Length: About 4 inches; the same size as the golden-crowned kinglet.

Male and Female: Similar in appearance to the house wren, but smaller and with a shorter tail; body brown, mostly barred with fine, black lines; light line over the eye; under parts darker than those of the house wren, with a buff wash across throat and breast.

Song: A very beautiful song, unusually loud for so small a bird. Those fortunate enough to hear it are extravagant in their praise. Mr. Eaton calls it the sweetest melody that he and his associates heard in the Adirondacks, excelling even the thrushes.

Habitat: Brush heaps, thickets in woods, along streams, and in wild rocky places.

Range: Breeds from southern Canada to Minnesota, Wisconsin, Michigan, New York, and Massachusetts, through the Alleghanies to North Carolina; winters from about its southern breeding limit to Texas and northern Florida.

Eaton says: "During the migration, this little wren is commonly observed about the shrubbery of our lawns, parks, and the edges of woods, when disturbed retreating to the recesses of some brush pile or under the damp edges of the stream bank. A few remain throughout the winter in western and central New York, and it is fairly common as a winter resident

[35] From Witmer Stone, in Educational Leaflet No. 50, National Association of Audubon Societies.

in the southeastern portion of the State, but in the principal breeding range 64 of the Adirondacks and Catskills it is only a summer resident."[36] It is a rather common winter visitor near Washington, and rare in New England.

THE DOWNY WOODPECKER

Woodpecker Family—Picidæ

Length: A little over 6½ inches; the smallest of our woodpeckers.

General Appearance: A *small* black and white bird, with a *white stripe extending down the middle of its back*; a red patch on back of male's head. The tail is used for a prop as the woodpecker climbs tree-trunks.

Male: Upper parts black and white; crown of head black with *red patch at nape*; two broad white stripes above and below eye; a *broad white stripe down the center of back*; wings spotted and barred with white; tail sharply pointed; the long tail-feathers, black; the short *outer tail-feathers, white barred with black*; bill long, strong, with a tuft of feathers at its base.

Female: Like male, except for the *absence of a red patch on the head.*

Notes: A call-note *Peek-peek.* A metallic *Tut-tut-tut'-tut-tut-tut-tut* might be considered the Downy's song, but he belongs really to the group of songless birds. He beats loud tattoos on the boughs of trees, especially at mating time.

Flight: Labored, jerky, with a characteristic shutting of the wings against the sides.

Habitat: Tree-trunks in woods and orchards, and on lawns. The Downy is our most common woodpecker, and a permanent resident.

Range: Northern and central parts of eastern North America, from Alberta, Manitoba, and Ungava, south to eastern Nebraska, Kansas, the Potomac Valley, and in the mountains to North Carolina.

[36] From Eaton's "Birds of New York."

Figure 11. Downy and hairy woodpeckers.

The southern downy woodpecker of the South Atlantic and Gulf States is smaller and browner than its northern relative.

The Downy Woodpecker is a member of a family of birds that has attracted man's attention since the old days of superstition. Various myths have grown up around these birds; those of the American Indians are possibly the most interesting. Until recently, woodpeckers have been persecuted by the white man, because of their habit of pecking at trees which they were thought to kill. Many have been unjustly slain.

While one branch of the family, the Sapsuckers, have done a great deal of harm to forests where they breed, and other woodpeckers have done occasional damage, it is now known that they are invaluable as preservers of our trees. Entomologists and foresters consider them the greatest enemies known of spruce-bark beetles and sap-wood borers. As borers are found near the surface in living trees, the holes made by woodpeckers while extracting them soon heals and leaves little mark.

An examination of the structure of woodpeckers shows the admirable way in which they are fitted for their work. They have short, stout legs; strong feet, usually with two toes in front and two in the back; large claws, and stiff tails tipped with sharp spines, to aid them in supporting themselves firmly against tree-trunks and branches. Mr. Forbush says: "The bird is thus more fully equipped for climbing than a telegraph lineman. The claws and tail take the place of the man's hand and spurs."[37]

Professor Beal writes the following:

> "As much of the 67 food of woodpeckers is obtained from solid wood, Nature has provided most of them with a stout beak having a chisel-shaped point which forms an exceedingly effective instrument. But the most peculiar and interesting point in the anatomy of these birds is the tongue. This is more or less cylindrical in form and usually very long. At the anterior end it generally terminates in a hard point, with more or less barbs upon the sides. Posteriorly the typical woodpecker tongue is extended in two long, slender filaments of the hyoid bone *which curl up around the back of the skull* and, while they commonly stop between the eyes, in some species they pass around the eye, but in others enter the right nasal opening and extend to the end of the beak. In this last case the tongue is practically twice the length of the head. Posteriorly this organ is inclosed in a muscular sheath by means of which it can be extruded from the mouth to a considerable length, and used as a most effective instrument for dislodging grubs or ants from their burrows in wood or bark. Hence, while most birds have to be content with such insects as they find on the surface or in open crevices, the woodpeckers devote their energies to those larvæ or grubs which are beneath the bark or even in the

[37] From "Useful Birds and Their Protection," by E. H. Forbush.

heart of the tree. They locate their hidden prey with great accuracy, and often cut small holes directly to the burrows of the grubs."[38]

Mr. Forbush calls attention to the wonderfully constructed head of a woodpecker

> "which is built so that it can withstand hard and continuous hammering. The skull is very thick and hard. Its connection with the beak is strong, but at the same time springy, and somewhat jar-deadening. 68 The membrane which surrounds the brain is very thick and strong."[39]

The Downy is the smallest member of the woodpecker family in North America, and is one of the most useful. He is especially fond of orchards and shade trees, and not only devours insects that infest them during the spring and summer, but eats the eggs they laid in the crevices of the bark during the winter. One Downy alone is of inestimable value in an orchard or a grove. Mr. Forbush writes as follows:

> "When the Metropolitan Park Commission first began to set out young trees along the parkways of Boston, some species of trees were attacked by borers; but the Downy Woodpeckers found them out and extracted the grubs, saving most of the trees.
>
> "The untiring industry of this bird and the perfection of its perceptive powers may be shown by the experience of Mr. Bailey. On March 28, 1899, a Downy Woodpecker that he watched climbed over and inspected one hundred and eighty-one woodland trees between 9:40 A. M. and 12:15 P. M., and made twenty-six excavations for food. Most of these holes exposed galleries in the trunks in high branches where wood-boring ants were hiding.... These ants often gain an entrance at some unprotected spot on a living tree, and so excavate the wood of the trunk that the tree is blown down by the wind. This woodpecker acts as a continual check on the increase of such ants."[40]

---

[38] Bulletin No. 37, U. S. Dept. of Agriculture, Biological Survey.

[39] From "Useful Birds and Their Protection," by E. H. Forbush, pages 245, 246, 252, 253.

[40] "Useful Birds and Their Protection," E. H. Forbush, pages 259 & 260.

The Downy may easily be attracted to our yards by a piece of suet fastened securely to a tree. During the past winter, one has sought my suet-cage, in company with chickadees and nuthatches. This spring he brought his 69 mate to a maple in front of the house. He has seemed excited and happy, and has drummed persistently on a certain broken limb of the tree. He has indulged in numerous rapid flights and his metallic, ringing call.

The Hairy Woodpecker

Woodpecker Family—Picidæ

Length: About 9½ inches; nearly ⅓ larger than the Downy, whom he *resembles almost identically as to general appearance, except in* SIZE.

Male: Black and white above; white underneath; *broad white stripe down* the *middle of* the *back*; head with black and white stripes, a red patch at the back, and bristles at the bill; wings black, with white stripes and bars; tail black, with white outside feathers; *the absence of black flecks on the tail-feathers and the larger size of the bird* distinguishes the Hairy from the Downy.

Female: Like male, except for the absence of a red patch on the head.

Note: A loud, shrill call, difficult to imitate or to reproduce on paper for identification. The Hairy also "drums" on the boughs of trees; it has no real song.

Habitat: Tree-trunks in woodlands, rather than in orchards or gardens, though I have noticed these woodpeckers in winter frequenting the trees of village streets without shyness or fear. During the breeding season, they remain in secluded spots in the woods.

Range: Three species of the Hairy Woodpecker may be found in Canada and the United States; the northern hairy woodpecker, the hairy woodpecker, and the southern hairy woodpecker. The northern species lives in the tree-zone of Canada, and is the largest of the three; the Hairy, next in size, may be found in the United States from Colorado, Nebraska, and Oklahoma, to the middle and northern parts of the Eastern States. The

Southern Hairy, the smallest of the three, is a resident of our southern section.

The Hairy Woodpecker is so like his small Downy relative in appearance and habits that his characteristics are not usually dwelt upon; he is like an older neglected cousin of a baby upon whom much attention is lavished.

But he is very worthwhile attracting. He is as untiring as the Downy in his quest for beetles, his favorite kind of tree-food; he is also a lover of ants and other "borers." His longer bill enables him to reach many that the Downy cannot. One Hairy Woodpecker alone saved an entire orchard that had become infested with "borers." One tree had died before he began his rescue-work, but he saved all the others.[40]

He likes the caterpillars of the cecropia and gypsy moths. He eats much vegetable food, especially during the winter; he has been known to take an occasional bite of the soft inner bark of trees and a drink of sap which he has well earned. Like the Downy, he will eat suet in the winter season. Mr. Forbush writes:

> "While this bird often excavates a hole for winter shelter, it sometimes sleeps exposed on a tree-trunk. Mr. Bailey and I once watched one that slept for many winter nights on the north side of a tree trunk in a thick grove. It attached its claws to the bark and went to sleep in much the same position in which it ordinarily climbed the tree. It invariably went to the same tree at night, and was found in the same place at daylight every morning."[40]

The White-Breasted Nuthatch

Nuthatch Family—Sittidæ

Length: About 6 inches.

General Appearance: A short, thickset bird, blue-gray, black, and white. Bill long; tail short and square.

Male: Mostly bluish-gray above; white underneath, shading to reddish-brown at sides and under tail; top of head and nape a shining blue-black; sides of head and throat white; wings gray shading to brown, edged and tipped with light gray or white; shoulders gray and black; bill large and strong, (¾ of an inch in length); tail short and square-cut; middle feathers bluish-gray; outer ones black, with large white patches near tips; legs short; feet large and strong; hind toe unusually long, with a long, sharp nail.

Female: Head a dull grayish-black; otherwise like male.

Notes: A nasal *crank-crank*, which, though not melodious, is not unpleasant to hear. Dr. Chapman says: "There is such a lack of sentiment in the Nuthatch's character, he seems so matter-of-fact in all his ways, that it is difficult to imagine him indulging in anything like song. But even he cannot withstand the conquering influences of spring, and at that season he raises his voice in a peculiar monotone—a tenor *hah-hah-hah-hah-hah*—sounding strangely like mirthless laughter."[41]

Flight: Undulating.

Habitat: Trunks of trees, which he ascends and descends. The other tree-trunk birds, except the black and white warbler, usually ascend trees.

Range: North America, east of the Plains. A permanent resident, though irregularly distributed. Breeds from central Canada to the northern parts of the Gulf States.

Of the so-called tree-trunk birds, none are easier to identify than nuthatches, because of their habit of *descending* trees. Woodpeckers jerk themselves up a tree somewhat as men might ascend telegraph-poles or smooth slippery palm trees. Creepers wind spirally about trunks in a gentle, unobtrusive manner. Both woodpeckers and creepers use their sharply-pointed tails as props. Not so the nuthatches. They care not how they go—"uphill or down dale"—all is one to them. They are as sure-footed as burros descending the Grand Canyon. If they depart from their trail, and decide to leap from crag to crag of their arboreal cliffs, they alight on their strong feet with something of the assurance of a cat. Their tails are not necessary to them as supports.

---

[41] From "Birds of Eastern North America," by Frank M. Chapman.

It is interesting to inquire into the reasons for curious habits of birds. In the economy of Nature one finds marvelous adaptations and harmonies. Mr. Francis H. Allen, in his delightful sketch written for the National Association of Audubon Societies, speaks of the nuthatch as "filling a gap in nature" by approaching his prey from an angle not possible to woodpeckers and creepers. Mr. Allen says:

> "He would not have adopted so unusual a method of feeding if it had not stood him in good stead. I suspect that by approaching his prey from above he detects insects and insect-eggs in the crevices of the bark which would be hidden from another point of view. The 74 woodpeckers and the creepers can take care of the rest. Of course these other birds get something of a downward view as they bend their heads forward, but the Nuthatch has the advantage of seeing, before he gets to them, some insects which even a Brown Creeper's gentle approach would scare into closer hiding in their holes and crannies."[42]

In addition to beetles, moths, caterpillars, ants, and wasps, the nuthatch eats seeds, waste grain, and nuts such as acorns, beechnuts, and chestnuts.[43] His habit of wedging nuts into some crevice that will hold them securely, and then using his strong bill as a hatchet to "hatch" open the nuts is well-known. From that habit he derives his name, which Mr. Forbush says originated probably from *nuthack* or *nuthacker*. The bird does much good, and no harm that is known.

He is active and cheerful, inquisitive, and intelligent. He makes an interesting winter companion. During an ice-storm in Asheville, N. C., a nuthatch was attracted by fragments of bread scattered for the hungry winter birds during their famine time. This nuthatch pounced on large crumbs so greedily and purloined them so rapidly that my sisters feared he would die of acute indigestion! They finally discovered that he had wedged the crumbs into large crevices in the bark of a tree near by, and had stowed one good-sized crust in a hole in a telegraph-pole. When he had

---

[42] Educational Leaflet No. 59.

[43] Farmers' Bulletin 513, U. S. Department of Agriculture, Bureau of Biological Survey.

appropriated most of the bread, he spent the day feasting, going from one store house to another.

A nuthatch in Massachusetts frequently sought an improvised 75 feeding-table made from a bluebird's nesting-box. One cold morning the owner saw him emerging from the box, where he had evidently "spent the night sitting on his breakfast," literally seated in the lap of luxury. He reminded me of that delicious tale I loved to read and contemplate during childhood,—of the children who lived in a candy house and ate their way out of it!

Another New England nuthatch, one that I watched at my feeding-table, at first made rapid inroads upon the suet-cage, storing pieces in the cracks of a tree nearby. I saw him tuck one large crumb beneath a warped shingle of the chicken-house, evidently laying it up for an icy day, instead of the proverbial rainy one. When an unusually severe ice-storm occurred, he returned to his store house and the crumb disappeared. I had the satisfaction of having assisted him in his dire need.

The Red-Breasted Nuthatch

Nuthatch Family—Sittidæ

The Red-breasted Nuthatch is very similar to its white-breasted cousin except that it is smaller, (4½ to 5 inches), and is yellowish or "rusty" underneath, (except for a white throat), has a white stripe on each side of its black crown, and a black stripe extending through the eye. The head of the female is gray, with white and gray stripes.

This species is not so well known as the white-breasted nuthatch, because it frequents coniferous forests or woods that contain evergreens. It breeds from the Upper Yukon Valley, central Canada, and northern United States, and winters as far south as lower California, New Mexico, Arizona, and the Gulf Coast.

Figure 12. Nuthatch.

Mr. Allen says of this bird:

"To those who know it the Red-breasted Nuthatch is dear out of all proportion to its size and its musical attainments. It is livelier than its big cousin, and prettier in its markings, and there is something particularly fetching about its quaint little form. It is even less of a songster than the white-breasted species, for prolongations and repetitions of its call-note seem to be all it has that can pass for a song. This call-note can be rendered as äap. It is nasal, like that of the White-breasted Nuthatch, but much higher in pitch, more drawling, and lacks the r. It has been happily likened to the sound of a tiny trumpet or tin horn.

"The habits of the Red-breasted Nuthatch are so like those of the White-breasted that much that I have said about that species is applicable to this. The most striking difference is in the favorite haunts of the two birds, the Red-breasted preferring the coniferous woods, or mixed woods that contain a large proportion of evergreens. In those winters when they are found in southern New England, they come freely to the neighborhood of man's dwellings and feed familiarly on the supplies provided for the winter birds, but even there they show their partiality for coniferous trees. They are particularly fond of the seeds of pines and spruces, so that they are much more vegetarian than their white-breasted cousins. They have the same habit of hiding their savings in cracks and crevices."[44]

The Brown Creeper

Creeper Family—Certhiidæ

Length: About 5½ inches.

Male and Female: Brown above, mottled with gray, buff, and white; under parts white. A *whitish line over eye*; bill long, curved; a bar of buff across wings; tail-feathers long, *sharply pointed*; *upper tail-coverts bright reddish-brown*.

Note: A faint, monotonous, *skreek-skreek, skreek-skreek*.

Song: According to Brewster, the brown creeper sings an unusually sweet song during the nesting season.

Habitat: Tree-trunks, which are carefully inspected by these industrious birds.

Range: Eastern North America. Breeds from Nebraska, Indiana, the mountains of North Carolina and Massachusetts north to southern Canada; also in the mountains of western North America from Alaska to Nicaragua; winters over most of its range.

---

[44] Educational Leaflet No. 59, National Association of Audubon Societies.

The Brown Creeper should inherit the earth, for he is one of the most perfect examples of meekness that may be found. Small, slight, self-effacing, untiring in his work, he reminds one of a quiet industrious person who performs unremittingly small tasks that amount to a large total.

He is a searcher for insect-eggs, and for insects so small that they might escape the notice of eyes not peculiarly fitted to espy them. His long bill is slender enough to slip into crevices which neither nuthatches nor woodpeckers investigate. Possibly it is because he selects such tiny particles of food that he must work so industriously in order to get enough to eat. He seems always in a hurry. Mr. Frank Chapman has humorously described the brown creeper as follows:

Figure 13. Brown creeper.

> "After watching him for several minutes, one becomes impressed with the fact that he has lost the only thing in the world he ever cared for, and that his one object in life is to find it. Ignoring you completely, with scarcely a pause, he winds his way in a preoccupied, near-sighted manner up a tree-trunk. Having finally reached the top of his spiral staircase, one might suppose he would rest long enough to survey his surroundings, but like a bit of loosened bark he drops off to the base of the nearest tree and resumes his never-ending task."[45]

The creeper is not easy to find. He is so wonderfully protected by his dull brown feathers that he looks more like an animated lichen than a bird. His nest is a cleverly camouflaged affair, tucked behind loose bark and often containing eight whitish eggs about the size of beans.

We are surprised to learn that this patient, hard-working little creature has the soul of a poet. His sweet nesting song, reserved for his mate brooding in the woods, breathes exquisite tenderness and beauty.

THE STARLING

Starling Family—Sturnidæ

Length: About 8½ inches.

General Appearance: A *short-tailed, long-billed black bird* with flecks of brown that look like freckles.

Male and Female: Head purple, flecked with light brown spots; body purple and green, the purple predominating on back and sides, the green on the breast. In summer, the upper parts and sides are speckled, the breast and belly dark, and the *bill yellow*. In winter, the upper parts are spotted with light brown, the under parts with white; the bill is *brown* until January, when it begins to turn yellow.

Notes: Squeaks and gurgles, interspersed with pleasant musical notes. A flock of starlings make a great deal of noise.

---

[45] From "Birds of Eastern North America," by Frank M. Chapman.

Range: Numerous starlings live in the Eastern Hemisphere. A number of them were brought to America in 1890 and released in Central Park, New York City. They have increased in number and enlarged their range greatly. They have spread northward and southward; they are now reasonably common near Boston and Washington, as well as New York and other places in the East.

In the winter, starlings are easily identified, because they are the only black birds smaller than crows to be found in some localities. In the spring, they may be readily distinguished from grackles because they have *yellow bills*, *dark eyes*, and *short, square* tails, while 81 grackles have *dark* bills, *yellow* eyes, and *long* tails. Both starlings and grackles are iridescent; a near view reveals the spotted plumage of the starlings and the iridescent bars on the backs of the purple grackles.

Major Bendire says that starlings possess unusual adaptability and can make their nests in a great variety of places. Accusations are brought against them for driving away bluebirds and even flickers. It remains to be seen how much harm is done to our native birds in this way.

There are different opinions regarding the economic value of Old World starlings. Mr. Forbush tells of an Australian locust invasion near Ballarat, Victoria, which made terrible havoc with crops. "It was feared that all the sheep would have to be sold for want of grass, when flocks of Starlings, Spoon-bills, and Cranes made their appearance and in a few days made so complete a destruction of the locusts that only about forty acres of grass were lost." Mr. Forbush gives also "the experience of the forest authorities in Bavaria during the great and destructive outbreak of the nun moth which occurred there from 1889 to 1891. The flight of Starlings collected in one locality alone was creditably estimated at ten thousand, all busily feeding on the caterpillars, pupæ and moths. The attraction of Starlings to such centers became so great that market-gardeners at a distance felt their absence seriously."[46]

In an article by E. R. Kalmbach of the Biological Survey, published in "The Auk" of April, 1922, and entitled "A Comparison of the Food Habits

---

[46] From "Useful Birds and Their Protection," by E. H. Forbush, pp. 65 and 17.

of British and American[82] Starlings," occur the following statements by Dr. Walter E. Collinge, the eminent Scotch biologist:

> "The Starling offers a most serious menace to the production of home-grown food, and any further increase in its numbers can only be fraught with the most serious consequences." He says also, "For many years past there has been taking place a sure but gradual change of opinion with reference to the economic status of the Starling, for from one of our most useful wild birds it has become one of the most injurious. Its alarming increase throughout the country threatens our cereal and fruit crops, and the magnitude of the plague is now fully realized." He states further, "There is fairly reasonable evidence to show that in the past the bulk of the food consisted of insects and insect larvæ, slugs, snails, earthworms, millepedes, weed seeds, and wild fruits; in more recent years, this has been supplemented by cereals and cultivated fruits and roots."

Mr. Kalmbach reports a better record for the starling in America, and refers to the decision made by the Department of Agriculture, reported in Bulletin 868:

> "Most of the Starling's food habits have been demonstrated to be either beneficial to man or of a neutral character. Furthermore, it has been found that the time the bird spends in destroying crops or in molesting other birds is extremely short compared with the endless hours it spends searching for insects or feeding on wild fruits. Nevertheless, no policy would be sound which would give the bird absolute protection and afford no relief to the farmer whose crops are threatened by a local overabundance of the species.... The individual farmer will be[83] well rewarded by allowing a reasonable number of Starlings to conduct their nesting operations on the farm. Later in the season a little vigilance will prevent these easily frightened birds from exacting an unfair toll for services rendered."

The Northern Shrike or Butcher-Bird

Shrike Family—Laniidæ

Length: A little over 10 inches.

Male and Female: Gray above, lighter underneath; forehead, rump, and upper tail-coverts white; wings black, irregularly marked with white; tail black, bordered with white; *a heavy black streak extending from the bill beyond the eye*; *bill hooked* and blackish.

Notes: A call-note and a sweet song.

Habitat: Fields or roadsides where it can find insects, small rodents, and little birds for its prey.

Range: Northern North America. Breeds from northwestern Alaska and northern Canada to the base of the Alaskan Peninsula, Saskatchewan, Ontario, and Quebec; winters south to central California, Arizona, New Mexico, Texas, Kentucky, and Virginia.

The loggerhead shrike, a resident of the Southern States, is similar to the Northern Shrike but smaller. It is found from southern Florida to North Carolina and west to Louisiana. Northward this species is represented by the migrant shrike, nesting locally from Virginia and eastern Kansas to the southern border of Canada.

Shrikes or Butcher-Birds are attractive to look at, but have a habit which renders them extremely unpopular. They pursue small rodents and little birds and impale them upon sharp twigs, thorns, or barbed wire fences. In excuse for these cruel acts, it must be said that they have not strong, sharp talons like hawks and owls; in order to tear their prey to pieces, there must be a way of holding 85 it firmly.[47] One agrees with Mr. Forbush, however, in his estimate of the habit. He says:

> "The Shrike or Butcher-Bird is regarded as beneficial; but our winter visitor, the Northern Shrike, kills many small birds. It pursues Tree Sparrows, Juncos, Song Sparrows, and Chickadees, overtakes and strikes

[47] From "Useful Birds and Their Protection," by E. H. Forbush, page 370.

them while they are in flight, sometimes eating them, but oftener leaving them to hang on trees, where they furnish food for other birds. When one sees the little Butcher killing Chickadees and hanging them up, his faith in its usefulness receives a great shock. Shrikes are probably of less value here than in their northern homes, where in summer they feed much on insects. Their chief utility while here [in Massachusetts] consists in their mouse-hunting proclivities."[47]

Their habit of killing English sparrows and thus getting rid of a nuisance has been commended. Shrikes are likewise destroyers of grasshoppers, crickets, beetles, and other insects.

"Like birds of prey and some other birds, the Butcher-Bird habitually disgorges the indigestible part of its food after digesting the nutritive portion. The bones and hair of mice are rolled into compact pellets in the stomach before being disgorged."[48]

[48] *Farmers' Bulletin 506*, U. S. Dept. of Agriculture, Biological Survey.

PART 2 - DESCRIPTIONS AND BIOGRAPHIES OF OUR EARLY SPRING BIRDS

Chapter 4

Introduction: Early Spring Birds

On a mild day late in February or early in March, before winter is really over and snow has entirely disappeared, one may hear the cheerful voice of the song sparrow, the welcome chirp of the robin, or the sweet note of the bluebird. Even though ice and snow return, courage is renewed with the advent of winged messengers who presage the ever fresh miracle of spring, and who hold home-love in their hearts so strong that they brave cold and distance to return to the "Land of Their Hearts' Desire."

As the season advances, other birds arrive. A "dusky line" of wild geese "honk" noisily; flocks of grackles "creak" from the pines; red-winged blackbirds join the hylas in awakening the marshes; phœbes call disconsolately for their mates; fox sparrows, chewinks, and white-throats sing melodiously from thickets; cowbirds appear in fields, which ring with the clear songs of meadowlarks and the tender notes of field and vesper sparrows. Mourning doves coo gently to each other; chipping sparrows make their homes in our gardens; kingfishers sound their rattles; flickers and red-headed woodpeckers raise their loud voices. The hills "clap their hands with joy"; the earth shows a flush of green and gold; trees and shrubs are touched with colors more exquisite than in autumn; wild-flowers carpet the woods and fields, and brooks join in the chorus of bird-song.

As the birds appear, it is not difficult to distinguish 90 them, if one begins before the great migration of late April or early May, and goes forth with alert senses and infinite patience and perseverance. With a reliable guide-book, a learner may be reasonably sure of the early migrants, because only certain species of large and confusing families are to be found during March and early April.

In watching birds, a student learns to observe with lightning speed; to note color and comparative size; distinguishing marks such as crests or striped crowns, spots on breast or throat, bars on wings or tail; the length and shape of bill, wings, tail, and legs. He learns also to notice whether the bird walks, runs, hops, or "teeters"; whether its flight is swift or slow, direct like a robin's, undulating like a goldfinch's, soaring like that of hawks and eagles, labored or jerky like woodpeckers', or graceful and "skimming" like that of swallows.

A careful observer notices also whether the bird was seen in a plowed field or a grassy pasture; by a roadside or in a thicket; in an orchard or an open grove; in deep woods or coniferous forests; in a treetop, on a tree-trunk, on the ground; near a stream, a pond, or a marsh; near a sandy or a rocky shore; in an arid region, or among mountains.

A sure means of identification for many species is the song or the call-note. The songs of some birds are similar to those of others, but there is usually a characteristic note or strain. When beginning my study of birds, I traced every sound I could to its source, waited till I saw the author of the note or song, listened till I learned it, could reproduce it, or at least be sure of future recognition. I found that the training of my sense of hearing opened a 91 avenue of enjoyment of which I had been utterly unconscious; many others testify to a similar pleasure. Thoreau speaks repeatedly of his joy in sound and even in silence. Truly the voice of God may thus be heard and His infinite power further revealed.

Chapter 5

Spring Migration Lists

Dates of Arrival of "Summer Visitants" Near New York City

February 15 to 28

- Purple Grackle
- Rusty Blackbird
- Red-winged Blackbird
- Robin
- Winter Residents and Visitants

Birds Seen in March

Winter Residents Leaving for the North

- Snowflake
- Northern Shrike

- Horned Lark
- Redpoll

Migrants Arriving from the South

- Loon
- 4 species of Ducks

March 1 to 10

- Purple Grackle
- Red-winged Blackbird
- Rusty Blackbird
- Robin

March 10 to 20

- Phœbe
- Meadowlark
- Cowbird
- Fox Sparrow
- Woodcock

March 20 to 31

- Kingfisher
- Mourning Dove
- Swamp Sparrow
- White-throated Sparrow
- Wilson's Snipe

Birds Seen in April

Winter Residents Leaving for the North

- Junco
- Tree Sparrow
- Winter Wren
- Brown Creeper
- Red-breasted Nuthatch
- Golden-crowned Kinglet

Migrants Arriving from the South

April 1 to 10

- Great Blue Heron
- Black-crowned Night Heron
- Osprey
- Vesper Sparrow
- Field Sparrow
- Chipping Sparrow
- Tree Swallow
- Myrtle Warbler
- Hermit Thrush

April 10 to 20

- Yellow-bellied Sapsucker
- Barn Swallow
- Yellow Palm Warbler
- Pine Warbler
- Louisiana Water-thrush
- Ruby-crowned Kinglet
- Green Heron

April 20 to 30

- Whip-poor-will
- Chimney Swift
- Least Flycatcher
- Towhee
- Purple Martin
- Cliff Swallow
- Bank Swallow
- Rough-winged Swallow
- Black and White Warbler
- Black-throated Green Warbler
- Brown Thrasher
- Spotted Sandpiper

Birds Arriving In May

May 1 to 10

- Yellow-billed Cuckoo
- Black-billed Cuckoo
- Nighthawk
- Ruby-throated Hummingbird
- Crested Flycatcher
- Kingbird
- Baltimore Oriole
- Bobolink
- Indigo Bunting
- Rose-breasted Grosbeak
- Scarlet Tanager
- Red-eyed Vireo

- Warbling Vireo
- Yellow-throated Vireo
- White-eyed Vireo
- Blue-winged Warbler
- Parula Warbler
- Black-throated Blue Warbler
- Magnolia Warbler
- Yellow-breasted Chat
- Chestnut-sided Warbler
- Hooded Warbler
- Yellow Warbler
- Maryland Yellow-throat
- Oven-bird
- Redstart
- House Wren
- Catbird
- Wood Thrush
- Veery

May 10 to 20

- Wood Pewee
- White-crowned Sparrow
- Golden-winged Warbler
- Worm-eating Warbler
- Blackburnian Warbler
- Bay-breasted Warbler
- Black-poll Warbler
- Wilson's Warbler
- Canadian Warbler
- Marsh Wrens
- Olive-backed Thrush

- Gray-cheeked Thrush
- Bicknell's Thrush

Summer Visitors That Breed Farther South and Are Occasionally Seen Near New York

- Red-bellied Woodpecker
- Summer Tanager
- Carolina Chickadee
- Blue-gray Gnatcatcher
- Mockingbird

Numerous Water-birds that nest in the Antarctic regions visit our shores during the summer.

Fall Migration

Summer Residents Leaving for the South

September 1 to 10

- Orchard Oriole
- Rough-winged Swallow
- Worm-eating Warbler
- Blue-winged Warbler

September 10 to 20

- Baltimore Oriole
- Purple Martin
- Yellow Warbler
- Yellow-breasted Chat

September 20 to 30

- Green Heron
- Hummingbird
- Kingbird
- Crested Flycatcher
- Wood Pewee
- Rose-breasted Grosbeak
- Yellow-throated Vireo
- Warbling Vireo
- Hooded Warbler
- Louisiana Water-thrush
- Veery

Migrants Arriving from the North

September 1 to 10

- Black-poll Warbler
- Connecticut Warbler

September 10 to 20

- Wilson's Snipe
- Olive-backed Thrush
- Bicknell's Thrush

September 20 to 30

- Herring Gull
- Junco
- White-throated Sparrow
- White-crowned Sparrow
- Myrtle Warbler

- Yellow Palm Warbler
- Brown Creeper
- Golden-crowned Kinglet
- Ruby-crowned Kinglet
- Winter Wren
- Gray-cheeked Thrush

October 1 to 10

- Black-crowned Night Heron
- Yellow-billed Cuckoo
- Black-billed Cuckoo
- Chimney Swift
- Least Flycatcher
- Bobolink
- Indigo Bunting
- Scarlet Tanager
- Cliff Swallow
- Barn Swallow
- Bank Swallow
- White-eyed Vireo
- Black and White Warbler
- Oven-bird
- Redstart
- Wood Thrush

October 10 to 20

- Spotted Sandpiper
- Whip-poor-will
- Nighthawk
- Red-eyed Vireo
- Maryland Yellow-throat

- Catbird
- Brown Thrasher
- House Wren
- Marsh Wren

October 20 to 31

- Phœbe
- Towhee
- Tree Swallow

Migrants Arriving from the North

October 1 to 10

- Bronzed Grackle
- Rusty Blackbird
- Hermit Thrush
- Canada Goose
- Loon
- Pintail and Mallard Ducks

October 10 to 20

- Fox Sparrow

October 20 to 31

- Horned Lark
- Tree Sparrow
- Snowflake
- Redpoll
- Northern Shrike

November

Migrants Leaving for the South

- Mourning Dove
- Belted Kingfisher
- Cowbird
- Red-winged Blackbird
- Purple Grackle
- Vesper Sparrow
- Chipping Sparrow
- Field Sparrow

Birds Seen in December

Permanent Residents, Winter Residents and Visitants

It is interesting to note that the earliest arrivals in the spring are the last to migrate in the fall. The reason is the food-supply. The insectivorous birds arrive later and leave earlier than those that have a more varied diet. An unusually severe winter sends birds south of their usual winter range.

The dates of migration must necessarily vary with latitude. 95 Migrants arrive near Washington a week or two earlier than near New York City, and near Boston a few days later. The lateness of the spring sometimes causes a delay of a week or two. The May arrivals appear more nearly on schedule. After May 15 birds begin to decrease in number, the "Transient Visitors" passing farther north; by June 5 we have with us our "Permanent Residents" and "Summer Residents."

In the fall the mildness of a season may cause November migrants to remain into December, or an open winter may tempt those that habitually migrate only a short distance to remain north of their usual winter range.

Chapter 6

Descriptions and Biographies of Early Spring Birds

The American Robin

Thrush Family—Turdidæ

Length: 10 inches.

Male: Head black; bill yellow; a white spot above and below eye; throat white, streaked with black; back and wings gray; tail black, with white spots near tips of outer feathers; white beneath tail; entire breast and sides reddish-brown; color less brilliant in autumn and winter, and bill darker.

Young Female: Paler than male.

Young: Similar to female, except for speckled breasts and backs.

Call-note: A sharp *tut*, used to express anger or alarm; also a sweet tender note, with which it encourages its young or converses with its mate.

Song: A loud, clear morning song, *Cheer-up, cheer-up, cheer-up, cheer-up*, sweeter and more subdued toward evening. The song varies decidedly with different individuals. Many robins seem to enjoy improvisations; we may hear them sing their somewhat monotonous strain with pleasing variations. During their sojourn in the South they sing but little, and live in flocks remote from human habitations; consequently they are not loved as they are in the North.

Range: North America, breeding from the tree-limit south to the northern part of the Gulf States and Mexican tableland; in winter, to Florida and the highlands of Guatemala.

Figure 14. Robin.

None of our birds is so well-known and so universally beloved as the robin. He, together with the song sparrow and the bluebird, arrives at a time when we are weary of winter and yearning for spring. He seems to show so much eagerness to return to us that he receives a hearty welcome. He is the first bird that we knew in childhood, unless it be the English sparrow; our earliest books were filled with tales and poems concerning him. Most of us have a fund of anecdotes that we could relate.

A robin has distinct individuality. His is a many-sided nature. He is cheerful and optimistic, aggressive and fearless, pugnacious and ardent—like the brave Lochinvar, "so daring in love and so dauntless in war,"—yet withal tender, joyous, and lovable. He is a fighter at mating time, but a gentle husband.

There are few bird-choruses as sweet as robins' rain-song or even-song. I recall a flock of these happy birds singing from maple-tops in a little village nestled beside a brawling river, when patches of brown earth showed beneath melting snow, and heavy rain-clouds broke away to reveal a golden western sky. The robins sang with the joy that my own heart felt at the renewal of life on the earth. I once heard their even-song in an elm-shaded college-town of Massachusetts during a lovely Sunday evening in June, when church-bells rang and robins held a vesper service all their own. My sister and I walked beneath the great arched trees and found ourselves speaking in whispers, as was our habit in the cathedrals of the Old World.

The robin's tut-tut, or tut-tut-tut′-tut-tut-tut-tut,—his scolding note,—is very similar to the exclamation of reproof our grandfather used to administer to us for childish misdemeanors. It is amusing to see how robins use this form of remonstrance to humans. John Burroughs wrote that he was kept out of his own summer-house by a 98 female robin that was nesting there. She scolded him so soundly for trespassing upon his own property, which she had appropriated, that he could have no peace. He finally left her in possession till her young had flown.[49] I had a similar experience when picking cherries in a friend's garden. A robin had preceded me and resented my intrusion in no uncertain manner. No angry fishmonger of Billingsgate ever hurled more noisy vituperation at a thief than did that robin fling at me, especially when I coolly refused to heed his commands to "Keep Off."

---

[49] From "Under the Maples," by John Burroughs, p. 55.

I recall an amusing experience with a robin family one summer. The second brood of hungry babies were clamoring for "More," and following their overworked father about as I have seen human babies tease their mothers. He was decidedly "frayed" as to temper, but he chose to assume the entire parental responsibility. His faded, bedraggled spouse, perched disconsolately upon the roof of the chicken-house, flew down two or three times into the bosom of the family and endeavored to "do her bit"; but her testy husband drove her off each time with a sharp tut-tut, until in despair she remained upon the ridge-pole peeping forlornly. The father proceeded to pull up worms for his gaping brood in a manner so irritated and strenuous that I wondered whether he had had a "family jar," or was only worn out with anxiety and overwork. It is a huge task to feed one baby robin alone, who can eat sixty-eight angleworms a day,[50] or one hundred and sixty-five cutworms.[51]

Robins do good to the soil by dragging forth earthworms and preventing their too rapid increase. Mr. Forbush calls attention to the value of these birds in devouring "dormant cutworms and caterpillars even in February," also quantities of the larvæ of March flies and white grubs that injure grass. The robin is an enemy of caterpillars, especially those that live near the ground; his destruction of cutworms and white grubs alone entitles him to our gratitude. He does eat early cherries, and has been bitterly arraigned for so doing. When later cherries, apples, peaches, pears, and grapes are ripe, wild fruits and mulberries which he eats by preference, have also matured; so on the whole, he does little harm.[52] He is now protected in most of our states.

A Maine robin that had an inordinate love for cherries and garden-raspberries was at first intimidated by a most lifelike, well-set-up scarecrow placed in the garden for his benefit. But he grew wiser as the days passed: he approached the fearful creature and received no harm. Familiarity finally bred contempt, for one day he was discovered perched upon the scarecrow's shoulder eating a raspberry!

[50] Prof. D. Treadwell.
[51] Mr. Chas. W. Nash.
[52] Farmers' Bulletin 630.

Robins become very tame. I once had the pleasure of the companionship of a dear, gentle, little English robin—a bird very different in size and manner from his American cousin—who would come out of the shrubbery whenever I called him. He would approach within two or three feet of my chair, to snatch the soft crumbs that I placed on the ground to lure him. He rewarded me frequently with his delightful little bubbling song.

An American robin during a March ice-storm learned 100 that bread crumbs were to be found upon the window-sill of a house in Cleveland. He flew to the sill frequently. When he found no crumb awaiting him, he would tap on the pane, then fly away a short distance and remain until a fresh supply appeared. He and his mate nested in an apple-tree nearby. They and their brood were fed in this way the entire season by their bird-loving friends, until they were in danger of becoming pauperized! One morning the following March while the Cleveland family were breakfasting, they heard the familiar tap upon the pane! There was Robin back again—you may imagine his welcome! For four years, he continued to announce his arrival in the same manner, and to build in the same yard; each year he and his family were supplied with part of their food by their devoted friends. Then ill must have befallen him, for he never returned.

To another Ohio woman came the joy of having a robin enter her room frequently. She had tempted him with crumbs inside a window-sill. One day he perched upon the sewing-machine where she was at work, and sang his sweet song to her, as the busy machine hummed its tune.

A robin's nest is an untidy affair, but it is something that we should miss were it not a part of our environment. Few birds' eggs are more lovely in color than those of the familiar robins'-egg blue, nestled in their grass-lined cup of clay. Olive Thorne Miller wrote of a clever robin that wished to build her nest during an almost rainless spring. She could find no mud, so she waded about in her drinking-dish to wet her legs; she then hopped into the dust, and with her bill scraped the mud off her legs. 101 This she did repeatedly, until she had the necessary amount.[53]

[53] From Olive Thorne Miller's "First Book of Birds."

I once saw a mother-robin sheltering her brood during a rainstorm of great violence. Her soft body and outspread wings were pelted by the rain, but she seemed quite oblivious to everything except to keep harm from her young. Her protecting attitude and the look in her bright eyes made as beautiful an expression of mother-love as I ever witnessed.

The Bluebird

Thrush Family—Turdidæ

Length: About 6½ to 7 inches.

General Appearance: Upper parts bright blue; under parts reddish-brown; *no crest.*

Male: Head, back, and tail bright blue; wings blue, edged with black; in the fall, edged with reddish-brown; throat, breast, and sides reddish-brown; white from center of breast to tail.

Female: Similar to male, but paler; wings and tail brightest in flight.

Young: Grayish-blue, speckled with whitish; wings and tail bluish.

Call-note: An indescribably sweet rendering of the syllables, *Cheer-e-o*, given usually while the bird is on the wing.

Song: A gentle warble of exceptional sweetness—*whew'-ee, whew'-ee, whew'-ee*, uttered tenderly and pensively.

Habitat: Orchards and gardens. The birds are usually seen in pairs, and like rather conspicuous perches, such as fence-posts and telegraph wires.

Nest: Made of grasses and placed in old hollow trees, preferably apple-trees. One objection raised against tree-surgery is that it deprives bluebirds of nesting-sites, but that objection may be removed by furnishing nesting-boxes.

Range: Eastern North America. Breeds from southern Canada and Newfoundland to the Gulf Coast and Florida, west to the Rockies; winters in the southern half of the eastern U. S., south to Guatemala.

Figure 15. Bluebird.

As spring approaches, I invariably "go a-hunting," not for "rabbit-skins," but for song sparrows and bluebirds. Robins usually seek us, and sometimes their blue-winged cousins call *Cheer-e-o* as they fly swiftly over our housetops; but I am never happy until I have visited an orchard or pasture frequented by these heaven-sent birds. "My heart leaps up when I behold" once more their exquisite blue and hear their soft, delightful warble. Then I know that spring is really on her way, and I am again eager and expectant.

Bluebirds have always been much beloved, especially in New England. Florence Merriam writes: "Although the Bluebird did not come over in the Mayflower, it is said that when the Pilgrim Fathers came to New England this bird was one of the first whose gentle warblings attracted their notice, and, from its resemblance to the beloved Robin Redbreast of their native land, they called it the Blue Robin."[54]

The bluebird has always been a favorite theme for poets and nature-writers, especially in New England, where the beauty and warm coloring of this sweet bird seem exceptionally welcome after a long, severe winter. In Thoreau's diary, "Early Spring in Massachusetts," he refers to the bluebird thirteen times and writes:

> "The bluebird—angel of the spring! Fair and innocent, yet the offspring of the earth. The color of the sky, above, and of the subsoil beneath, suggesting what sweet and innocent melody, terrestrial melody, may have its birthplace between the sky and the ground."[55]

Burroughs, too, makes frequent mention of the bluebird. 104 In "Under The Maples" he says:

> "None of our familiar birds endear themselves to us more than does the bluebird. The first bluebird in the spring is as welcome as the blue sky itself. The season seems softened and tempered as soon as we hear his note and see his warm breast and azure wing. His gentle manners, his soft, appealing voice, not less than his pleasing hues, seem born of the bright and genial skies. He is the spirit of April days incarnated in a bird. Not strictly a songster, yet his every note and call is from out the soul of harmony."[56]

Bluebirds are of economic as well as æsthetic value. They devour cutworms and other kinds of caterpillars, grasshoppers, crickets, katydids, and beetles. They eat fruit in the winter; they prefer that taken from

---

[54] From "Birds of Village and Field," by Florence Merriam.
[55] Used with permission of the Houghton Mifflin Co., the authorized publishers.
[56] Used with permission of the Houghton Mifflin Co., the authorized publishers.

pastures, swamps, and hedgerows, rather than from gardens or orchards. They never destroy cultivated crops; on the contrary, benefit them.[57]

These birds are such devoted lovers that one is rarely seen far from its mate. The female is very gentle and timid; she seems to need reassurance and protection. There are times, however, when she knows her own mind and shows firmness of character. A male bluebird in Asheville, N. C., intoxicated by the warmth of a sunshiny January day, wooed a female ardently. She was very distant and finally dismissed him. She evidently had sufficient foresight to realize that it would be disastrous to go to housekeeping so early and therefore withheld her consent.

Numerous instances have been recorded of bluebirds that have lost their mates by accident and have mourned so deeply as to touch the heart of anyone who saw the tragedy or heard the cries of sorrow.

The Song Sparrow

Finch Family—Fringillidæ

Length: A little over 6 inches; about the size of the English sparrow.

General Appearance: A small brown bird with a grayish breast, a body heavily streaked with black, *a black spot in the center of breast*, and at each side of the throat.

Male and Female: Brown head with black streaks, a grayish line in center and over eye; brown line back of eye; back brown and gray, streaked with black; wings brown, with black spots,—no white bars; throat grayish-white; *a dark patch on each* side of throat; *a conspicuous black spot in center of breast*; belly white; sides whitish, streaked with brown and black; tail long, brown, darkest in center.

Call-note: *Chip, chip*—sharp and metallic.

Song: A sweet cheerful strain, with considerable variety in different individuals. It usually consists of three notes that sound like "See? See?

---

[57] Farmers' Bulletin 755, Biological Survey, U. S. Department of Agriculture.

See?" followed by a short trill. Henry van Dyke interprets the song as *Sweet, sweet, sweet, very merry cheer.*

Habitat: Bushes; near water, preferably.

Range: North America, east of the Rocky Mts. Breeds in Canada from Great Slave Lake to Cape Breton Island, south to southern Nebraska, central Missouri, Kentucky, southern Virginia, the mountains of North Carolina. Winters from Nebraska, Illinois, Massachusetts (locally) and New Jersey, south to the Gulf Coast.

The Song Sparrow, like air and sunshine, is a part of our daily lives after we have once become acquainted with him. In some localities he takes up his abode permanently; in others, he arrives in late February or early March and remains until November. Joy in life and deep contentment abide with him. He is the most incurable optimist of my acquaintance. I have heard him sing beside a brook that has just broken its icy fetters, while patches of snow still remained on the ground; during days of rain which silenced most songsters; through hot summer noons and during the almost songless molting-season,—nothing seems to daunt him, from early morning until sunset. Occasionally during the night is heard his simple strain, as though he needs must sing in his sleep.

His song is pleasing, but in no way remarkable. It is in a major key and lacks the ecstasy and piercing sweetness of the fox sparrow's, and the exquisite tenderness of the field and the vesper sparrow's, but it possesses a charm all its own. It breathes a joy in simple things—a steadfast and cheerful courage that makes us say, "He, too, is no mean preacher."

Song sparrows, like other members of the Finch family, are of great service in their destruction of insects and weed seeds, of which they consume enormous quantities. They eat wild berries and fruits only when their favorite food is not obtainable. They possess no bad habits and are desirable "bird-neighbors" to cultivate. Water always attracts them; one is most likely to find them near streams, in which they love to bathe.

Their nests are made largely of grasses, dead leaves, and root-fibres, and are lined with soft grasses. They 108 are placed in bushes or on the ground. The eggs, pale in color and flecked with brown, are well concealed by their markings. Song sparrows, usually serene, grow intensely nervous when the nest is approached, and betray its whereabouts by their incessant *Chip, chip*.

Figure 16. Song sparrow.

The Song Sparrow

"See? See? See? The herald of spring you see!
What matters if winds blow piercingly!
The brook, long ice-bound, struggles through
Its glistening fetters, and murmurs anew
With joy at the freedom the days will bring
When the snow has gone! And I, too, sing!
"See? See? See? A flush of color you see!

The tassels are hung on the budding tree,
Before it has drawn its curtain of leaves
To shade the homes of the birds. Now weaves
The silent spring a carpet fair,
With wind-flower and hepatica there.
"See? See? See? You are glad to welcome me.
You will hear my voice ring cheerfully
Through Summer's heat or days of rain
Until the winter has come again.
From dawn till dusk, my heart is gay,
And I sing my happy life away.
See? See? See?"

THE FOX SPARROW

Finch Family—Fringillidæ

Length: A little over 7 inches; about an inch longer than the English sparrow, and nearly as large as a hermit thrush.

Male and Female: Upper parts reddish-brown, *brightest* on *lower back* and *tail*. (The red-brown tail is a distinguishing mark of the fox sparrow as it is of the hermit thrush.) Under parts grayish-white; *throat*, *breast*, *belly*, and *sides heavily* and *irregularly streaked* with reddish-brown and black, except the middle of the belly, which is white.

Note: A faint *seep* or *cheep*.

Song: The most beautiful of all the sparrows'—a burst of melody possessing sweetness and power; joyous, yet with a minor strain.

Habitat: Tall thickets or clumps of weeds.

Range: North America. Breeds in the forest-regions of Canada and Alaska; winters from the lower Ohio and Potomac Valleys to central Texas and northern Florida.

Never shall I forget the thrill of surprise and ecstasy which my first fox sparrow brought to me! My sister and I were on eager quest for early migrants in open woods and overgrown pastures, when from a thicket of tall shrubs there burst so marvelous a “concord of sweet sounds” that we were spell-bound. No words can describe the tenderness, the joyous abandon, yet withal the strain of sadness in the song, as though the choristers had drunk deep of life, had visioned clearly its secrets, 110 and transmuted its experiences. When the music had become a soft cadence, we sought the singers, and found a band of thrush like sparrows scratching in the old brown leaves like bantam hens. They remained in the thicket for several days, singing most rapturously toward sunset.

Though shy birds and seen infrequently, fox sparrows occasionally approach houses. During a deep spring snow that covered the birds’ natural food-supply, several of these north-bound migrants came three times a day with a flock of juncos to feed on bread-crumbs in our back yard. Like Tommy Tucker, they “sang for their supper.” Twice they arrived before a fresh supply of crumbs had been scattered; their songs announced their presence and were accompanied by the gentle trill of the juncos. A large flock remained in Middlesex Fells for several days.

Most bird-lovers consider an experience with fox sparrows as out of the ordinary. Thoreau wrote: “Is not the coming of the fox-colored sparrow something more earnest and significant than I have dreamed of? These migrating sparrows bear all messages that concern my life.”[58]

THE PHŒBE

Flycatcher Family—Tyrranidæ

Length: About 7 inches; a little larger than the English sparrow.

Male and Female: Grayish-brown above; under parts light gray with yellowish wash; breast darker than throat, sides grayish-brown; head dark brown, somewhat crested; bill black, slightly hooked at tip, with bristles at base; wings dark brown, *with inconspicuous whitish wing-bars*; tail dark brown; edge of two outer tail-feathers yellowish-white.

---

[58] “Notes on New England Birds,” Thoreau, p. 311.

Song: No real song. Flycatchers are songless birds. The note is a hoarse *Phœbe*, sometimes *Pe-wit-Phœbe*. It is usually uttered mournfully and monotonously; occasionally the male gives numerous *Phœbes* rapidly while on the wing.

Figure 17. Phœbe.

Habitat: Near streams preferably. A favorite nesting site is underneath a bridge; eaves of barns or beams of piazzas are also used.

Range: Eastern North America. Breeds from north-central Canada south to northeastern New Mexico, central Texas, northern Mississippi and highlands of Georgia; winters south of latitude 37° to southern Mexico.

When March has lost some of its bluster and gentler weather prevails, there arrives from the land of sunshine and teeming insect life, a small brown and gray bird—the Phœbe, first of the Flycatcher family to come North. Like many of the early migrants, he travels without his beloved little mate, whom he seems to 112 miss sadly; for he sits disconsolately on a bare twig and calls her name in hoarse, wheezy tones. After she appears, it is pleasant to see their devotion, not only to each other, but to the nesting site. How they journey apart the great distance from South to North and find their own especial bridge or barn year after year, is one of the great mysteries.

Their large, loosely-constructed nest is made of moss and mud, lined with soft grass, hair, or feathers. It is usually infested with bird-lice, as I discovered, to my dismay. It is well not to allow phœbes to build where the lice may become a nuisance.

Like all the soberly-dressed flycatchers, phœbes seek conspicuous perches such as posts or dead branches. They have the family habit of ruffling up their head-feathers into a sort of crest, and of jerking their tails frequently, especially when uttering their note. They make unexpected sallies after insects, which their unusually keen eyes can see from dawn until dark.

Phœbes are among our most useful birds, for they destroy injurious beetles, weevils, flies that annoy cattle and horses, house flies, ants, mosquitoes, wasps, spiders, grasshoppers, and numerous other harmful insects.[59] Their soft brown and gray plumage blends with dull March meadows, with the silver sheen of the brooks they love, and with silken pussy-willows and brown willow-boughs.

The Black Phœbe

The Black Phœbe is found from Texas west to the Pacific coast. It catches flies persistently and well deserves its family name. In appearance

[59] Farmers' Bulletin 630, U. S. Dept. of Agriculture, Biological Survey.

it resembles the slate-colored 113 junco, for it has a dusky head, back, wings, tail, and breast, with a white belly. Professor Beal writes of this bird as follows: "The black phœbe has the same habits as its eastern relative, both as to selection of food and nesting sites, preferring for the latter purpose some structure of man, as a shed, or, better still a bridge over a stream of water, and the preference of the black phœbe for the vicinity of water is very pronounced. One may always be found at a stream or pool and often at a watering-trough by the roadside.

"Careful study of the habits of the bird shows that it obtains a large portion of its food about wet places. While camping beside a stream in California the writer took some pains to observe the habits of the black phœbe. The nesting season was over, and the birds had nothing to do but eat. This they appeared to be doing all the time. When first observed in the morning, at the first glimmer of daylight, a phœbe was always found flitting from rock to rock, although it was so dusky that the bird could hardly be seen. This activity was kept up all day. Even in the evening, when it was so dark that notes were written by the aid of the camp fire, the phœbe was still engaged in its work of collecting, though it was difficult to understand how it could catch insects when there was scarcely light enough to see the bird. Exploration of the stream showed that every portion of it was patrolled by a phœbe, that each one apparently did not range over more than twelve or thirteen rods of water, and that sometimes two or three were in close proximity."[60]

The Crow Blackbird or Purple Grackle

American Blackbird Family—Icteridæ

Length: 12 to 13½ inches. Tail about 5 inches long, nearly the length of that of the blue jay.

---

[60] Farmers' Bulletin 630, U. S. Dept. of Agriculture, Biological Survey.

General Appearance: A glossy black bird with *yellow eyes*, and a *long tail* that in flight resembles a pointed fan curving toward the midrib. Blackbirds *walk* instead of *hopping*.

Male: Black with beautiful iridescence; head, neck, throat, and breast with green, blue, and purple reflections; back and rump purple and green, with iridescent bars; wings and tail purplish; under parts duller.

Female: Duller than male, with less iridescence.

Call-note: A hoarse, loud *Chack*.

Song: A disagreeable grating noise that Mr. Forbush likens to the "rather musical creaking of a rusty hinge." I once noticed the strong resemblance of the sound to the squeaking wheels of farm-wagons that passed near a noisy flock of grackles. Blackbirds always look unhappy and uncomfortable when making their attempt at singing, as though they emitted the sound with great difficulty.

Habitat: Groves of pine and spruce, as dark and gloomy as the birds themselves. They are found in parks and meadows, on lawns and near buildings. They live in large flocks except at nesting time.

Range: Middle Atlantic coast-region of the United States. Breed from north shore of Long Island Sound (rarely in Massachusetts), the middle Hudson Valley west to the Alleghanies, and south to the uplands of Georgia, Alabama, and eastern Tennessee; winter mainly south of the Delaware Valley. The Bronzed and Florida grackles extend the range over the whole of eastern North America, to Great Slave Lake, Newfoundland, Colorado, and Florida.

It seems incredible that blackbirds should belong to the same family as sweet-voiced meadowlarks, gay bobolinks, and musical orioles. They are literally the "black sheep" of the family, with a plumage in keeping with their dark deeds, and a sinister expression that arouses suspicion and wins them few friends. Their habit of destroying birds' eggs and young birds makes them a terror to their neighbors. Dr. Frank Chapman humorously says that he "can imagine bird-mothers frightening their young into obedience by threatened visits from that ogre, the Grackle."[61] I saw a flock

[61] From "Birds of Eastern North America," by Frank M. Chapman.

of them invading the seclusion of Wade Park, Cleveland, one spring morning. Two irate robins drove three bandit blackbirds away from their nest with loud cries and swift pursuit. A few minutes later, I saw a wood thrush attack a grackle. She administered a severe blow upon his shoulder, which disarranged his feathers and left him in such evident pain as to be quite oblivious of my proximity. This habit of devastating nests is not, however, so general as has been supposed, for Professor Beal reports that "remains of birds and birds' eggs amount to less than half of one per cent. of his diet."[62]

Figure 18. Purple Grackle.

[62] Farmers' Bulletin 630, U. S. Department of Agriculture, Biological Survey.

During the breeding season, grackles do much good by their destruction of insects upon which their young are almost wholly fed. They devour beetles, the caterpillars 116 of gypsy and brown-tail moths, cutworms, grasshoppers, and locusts in great numbers. They "follow the plow" in search of the grubs and worms to be found in the up-turned earth.

Grackles are in great disfavor, however, because of the grain they consume. Professor Beal states that grain is eaten during the entire year except for a short time in the summer. Waste kernels are consumed during winter and early spring, but that eaten in July and August is probably standing grain. Middle-western farmers suffer considerably.[63] interesting to see blackbirds migrate. They fly in flocks thousands strong. Mr. Forbush tells of a flock which formed a black "rainbow of birds" that stretched from one side of the horizon to the other. There seemed to be "millions" of them. They fly with wonderful precision, like a well-trained army bent on destruction. They are truly "Birds of a feather" that "flock together" with a kind of joyless loyalty, disliked by most of the world.

The Bronzed Grackle

The Bronzed Grackle, like the Purple Grackle, has a purple head, but has a bronzed back without iridescent bars. It is found in central and eastern North America from Great Slave Lake to Newfoundland in Canada, south to Montana and Colorado, (east of the Rockies), and southeast to the northern part of the Gulf States, western Pennsylvania, New York, and Massachusetts. It winters mainly from the Ohio Valley to southern Texas.

The Florida Grackle

The Florida Grackle is abundant from South Carolina to Florida, and west along the Gulf Coast to southeastern Texas. It is similar to the Purple

---

[63] Farmers' Bulletin 630, U. S. Department of Agriculture, Biological Survey.

Grackle in appearance, but is smaller in size. Flocks of these grackles frequent groves of palmettoes and live-oaks.

The Boat-Tailed Grackle

The Boat-tailed Grackle, the largest member of the blackbird family, (16 inches long), has wonderful violet reflections on head and neck. The female is much smaller and is brownish. This grackle is found in the South Atlantic and Gulf States from Chesapeake Bay to Florida and west to the eastern coast of Texas, and like the red-winged blackbird seems to prefer the vicinity of water.

The Red-Winged Blackbird

American Blackbird Family—Icteridæ

Length: About 9½ inches; length varies in different individuals.

Male: Jet black, except shoulders, which are *scarlet* edged with *yellow*; plumage mottled in winter—upper parts edged with rusty brown; bill long, sharp-pointed, black; legs and feet black; eyes dark.

Female: Head and back blackish, rusty brown, and buff. Light streak over and under eye; throat yellowish; under parts streaked with black and white; wings brown, edged with buff; tail brown. Plumage inconspicuous, but attractive on close inspection.

Young Males: Similar to females, but with red and black shoulders.

Call-Note: A hoarse *chuck* resembling that of the grackle.

Song: A liquid, pleasant *o-ka-ree*.

Habitat:

In meadows where a streamlet flows
Or sedges rim a pool,
There swings upon a blade of green

Beside the waters cool,
A bird of black, with "epaulets"
Of red and gold. With glee
He plays upon his "Magic Flute":
"*O-o-ka-ree? O-o-ka-ree?*"

Nest: A beautiful structure, long and deep, fastened to reeds; a "hanging" nest.

Eggs: Pale bluish, with inky scrawls and spots.

Range: North America, east of the Great Plains, except the Gulf Coast and Florida; abundant where there are marshes and ponds; winters mainly south of Ohio and Delaware Valleys.

Figure 19. Red-winged blackbird.

When the hylas begin to pipe in the spring, they are joined by the musical Redwings. The voices of these birds have been likened to flutes, also to violincellos in an orchestra. Their song is pleasant to hear, but seems to require considerable effort on the part of the performers—they lift their shoulders and spread their tails into broad fans when singing.

Redwings are noisy chatterers; they are intensely social in their nature. It is thought that some males have several wives at a time—one marvels at their courage! During the winter the females flock by themselves, and in the spring migrate about two weeks after their venturesome, prospective husbands have come northward. When they arrive, there is great "Confusion of Tongues"—the marsh is transformed into a Babel. Then sites for homes are selected, and house-building begins in earnest. Blackbirds make devoted parents.

They are much more popular than their cousins, the grackles, though in some localities where they are very abundant, as in the Upper Mississippi Valley, they are in disfavor because of the grain they devour. They eat oats, corn, and wheat, but only one-third as much as do the grackles; they eat the seeds of smartweed and barnyard grass in preference. Grasshoppers they consider great delicacies, also many other harmful insects.[64] Professor Beal states that nearly seven-eighths of their food consists of weed seed and insects injurious to agriculture. He pleads for their protection as does Mr. Forbush, who says: "Should there be an outbreak of cankerworms in an orchard, the blackbirds will fly at least half a mile to 120 get them for their young."[65] They eat little fruit and do slight harm to garden or orchard. On the whole, they are beneficial to mankind.

The rusty blackbird and the yellow-headed blackbird are two other species of blackbirds.

The rusty blackbird resembles both the purple grackle and the redwing. It is more nearly uniformly glossy black in summer than the former; it is rusty in winter like the latter. It is about the size of the redwing and has a sweeter voice. It is sometimes mistaken for the grackle; but its smaller size, its shorter, rounder tail, and more musical voice differentiate it.

---

[64] Farmers' Bulletin 630, U. S. Department of Agriculture, Biological Survey.

[65] "Useful Birds and Their Protection," by E. H. Forbush, page 320.

The yellow-headed blackbird, our western species, is easy to identify because of his yellow head, neck, throat, and breast, and his black body, with white wing-patches. The female has a paler yellow head, which, with the breast, is marked with white.

The Yellowhead lives in swamps of the Mississippi Valley from Indiana westward to California. He is attractive to see, but not pleasant to hear. He, too, is a grain-thief and therefore unpopular.

Figure 20. Cowbird.

The Cowbird

American Blackbird Family—Icteridæ

Length: About 8 inches.

Male: Glossy black, with a brown head, neck, and breast; some metallic reflections on body, tail, and upper wing-feathers. *Smaller than the grackle*, with a shorter tail, less iridescence, and *dark eyes*. Like the grackle, the cowbird is a walker.

Female: Dark brown, with a grayish tinge; under parts lighter, especially the throat, which has two dark streaks outlining the light patch.

Call-note: A loud *chuck*.

Song: No real song, only a disagreeable gurgle, that is emitted with great effort.

Habitat: Pastures and open woodlands; usually seen on the ground, but sometimes in trees.

Range: North America. Breeds in central Canada, south to northern California, Nevada, northern New Mexico, Texas, Louisiana, and North Carolina; winters from southeast California and the Ohio and Potomac Valleys to the Gulf Coast and Central Mexico.

The four common black birds—crows, grackles, redwings, and cowbirds—all have sins laid at their doors. Crows and blackbirds are grain-thieves and destroyers of the eggs and young of other birds; redwings have been accused of polygamy and theft; but if judged by human standards, none compare with cowbirds in what might be called moral degeneracy. Cowbirds not only mate promiscuously, but unlike blackbirds, have no regard 122 for their own young. They are like the human mothers who lay their babies on doorsteps, depart, and let others rear them.

It is a well-known fact that the female cowbird always selects the nest of a bird smaller and weaker than herself in which to deposit her egg. Major Bendire lists ninety-one varieties of birds that have been thus outraged, frequent victims being the song sparrow, indigo bunting, parula warbler, yellow warbler, vireo, chipping sparrow, towhee, oven-bird, yellow-breasted chat, and even the tiny blue-gray gnatcatcher. From one to seven cowbirds' eggs have been found at a time in other birds' nests, often in the warm center of the nest. Unless the little bird should build a new floor, or abandon her nest entirely, the cowbird egg will hatch first, and the lusty changeling will demand the lion's share of food and attention. Frequently the other eggs do not hatch; if they do, the young birds often perish with hunger and cold. When young cowbirds have been reared by their patient little foster-parents, they leave their benefactors and join flocks of their disreputable relatives.

In justice it must be said that cowbirds, like all villains, have a redeeming trait—they are great destroyers of weed seeds and insects. Like Cadmus and his band, they "Follow the Cow," and enjoy the insects that she arouses as she walks about in pastures. When the cow lies down, they, too, pause; they have been known to hop upon her back in friendly fashion. Self-interest prompts them, however, for they know that they may find there a harvest of insects.

THE MEADOWLARK CALLED ALSO FIELD LARK AND OLD FIELD LARK

American Blackbird Family—Icteridæ

Length: About 10¾ inches, a little larger than the robin; bill 1½ inches.

General Appearance: A large brown bird, with a *short tail that shows conspicuous white feathers* at each side in flight. The bright yellow breast crossed by a black crescent is less frequently seen.

Male and Female: Upper parts dark brown, mottled with black and buff; head striped, with a light line through the center and a yellow line over each eye, alternating with two dark stripes; cheeks gray; throat, breast, and belly yellow; a V-shaped band on breast; sides and lower part of belly whitish, streaked with black; bill long and sharp; tail short, (about 3 inches); outer tail-feathers almost entirely white; middle feathers brown, barred with black.

Call-note: A sharp nasal *Yerk*, and a twitter that sounds like a succession of rapid sneezes.

Song: A loud, clear, sweet refrain that usually consists of four syllables, but sometimes of five or six. It has been interpreted in various ways as follows:

Spring′-of—the-y-e′-a-r!
I love—you d-e-a-r.
I'm Mead′-ow-lar′-rk.

Mr. T. Gilbert Pearson tells of a lazy darky down South who interpreted the lark's song as

"*Laziness-will kill′ you.*"[66]

Flight: Direct, yet fluttering; usually away from the observer, showing the brown back and white tail-feathers, as though the bird was conscious of its bright yellow breast.

Habitat: Cultivated meadows, and grassgrown fields, especially one containing a running brook for drinking and bathing. Its fondness for unmown fields has given it the name of "Old Field Lark."[67]

Range: Eastern North America. Breeds from eastern Minnesota and southern Canada, south to northern Texas, Missouri, and North Carolina, and west to western Iowa, eastern Kansas, and northwestern Texas; winters regularly from southern New England and Ohio valley south to the Gulf States, and north locally to the Great Lakes and southern Maine.

In the South, from southern Illinois, southwestern Indiana and North Carolina to the coast of Texas, Louisiana, and southern Florida is found the southern meadowlark, smaller and darker than the northern species, and with a different song.

---

[66] From Educational Leaflet No. 3, National Association of Audubon Societies.
[67] Farmers' Bulletin 755, Biological Survey, U. S. Department of Agriculture.

Figure 21. Meadowlark.

In the West, from British Columbia to Manitoba and south to southern California, northern Mexico, and Texas is the western meadowlark, similar to its eastern relative in habits and plumage, but very different as to song. Its pure, sweet, liquid notes are among my most delightful memories of western birds.

It is fortunate that no human being or bird is possessed of all the virtues and charms, and that every individual may hold his own place in our interest and affections. As the spring migrants arrive, each receives a welcome peculiarly his own.

> "The lark is so brimful of gladness and love—
> The green fields below him, the blue sky above,

That he sings and he sings and forever sings he,
'I love my love, and my love loves me.'"[68]

His voice, clear and sweet, rings out joyously across the fields, fragrant with up-turned earth and bright with sunshine. He is the delight of spring meadows as Bob White is of summer fields.

The meadowlark has many friends: those who love him for his winning ways—his brightness, cheerfulness, and devotion to his family; epicures, ignorant of his value or fond only of their own pleasure; and people who realize that he is of enormous economic importance.

He was formerly believed to be a destroyer of grain. He was accused of pulling up as much corn and oats as crows, and of eating clover seed; but he is now recognized as "one of the most useful allies of agriculture, standing almost without a peer as a destroyer of noxious insects."[69]

So untiring is he in his search, that he uses his long sharp bill, even while snow is on the ground, to probe the earth for larvæ. He rids the fields of grasshoppers, crickets, beetles, caterpillars, flies, spiders, and "thousand-legs." Grasshoppers are his favorite delicacy. Professor Beal states that these insects form three-fourths of the meadowlark's food during August. He eats also large numbers of the white grubs of beetles "which are among the worst enemies of many cultivated crops, notably grasses and grains, and to a less extent of strawberries and garden vegetables."[69]

Like the quail, meadowlarks destroy weed seeds, which are eaten mostly in winter. When insects are obtainable, they are greatly preferred.

A search for a meadowlark's nest is an exciting adventure that keeps one alert. It is usually found by accident, perhaps after the wary builder has ceased trying to deceive the searcher. A sight of the speckled eggs or young fledglings in their cozy home with a grass-arched doorway is not soon forgotten.

Unlike quail, baby meadowlarks are unable to run about as soon as they are out of the egg, but remain for two weeks in their cleverly

[68] Written by Coleridge about the European skylark, but applicable to our meadowlark.
[69] Farmers' Bulletin 630 and 755, U. S. Department of Agriculture, Biological Survey.

camouflaged home, where they are often the prey of snakes and other enemies. Meadowlarks are now being widely protected, for many farmers regard them as one of their greatest assets.

THE NORTHERN FLICKER OR GOLDEN-WINGED WOODPECKER

Woodpecker Family—Picidæ

Length: About 12 inches; one of our largest common birds.

General Appearance: A large brown bird with a *red patch on the back* of the *head*, *conspicuous white rump* and *yellow lining* of *wings*, which distinguish it from the brown meadowlark with its white tail-feathers.

Male: Top of head and neck gray; a *crescent* of *red across nape*; cheeks and throat pinkish-brown, separated by *black patches*; strong bill 1½ inches long; under parts pinkish-brown and white, *heavily spotted with black*; a *black crescent* separates throat and breast. Back and upper wing-feathers a grayish-brown, *barred with black*; large white patch at rump very conspicuous in flight; upper tail-coverts black and white; tail black above, yellow underneath.

Female: Like male, except for the *absence of black patches at the sides of the throat*.

Notes: A loud *che-ack'*; also a note which Mr. Frank M. Chapman says "can be closely imitated by the swishing of a willow-wand: *weechew, weechew, weechew*."[70] Flickers drum frequently on boughs, also, and give a loud, rapid *flick, flick, flick, flick, flick, flick, flick, flick, flicker*,—which may be called, by courtesy, their song.

Habitat: Open woods, fields, orchards, and gardens, where trees or ant-hills are to be found.

Range: Northern and eastern North America. Breeds in the forested regions of Alaska and Canada; in the United 128 States east of the Rockies

[70] From "Birds of Eastern North America," by Frank M. Chapman.

and southward to the Gulf Coast and Texas in the winter. Resident in the U. S. except in the more northern parts.

Figure 22. Flicker.

The southern flicker, a resident as far south as southern Florida and central Texas, is *smaller* and *darker* than the Northern Flicker.

The red-shafted flicker, a western species, has *red cheek-patches* instead of *black*, *red wing* and *tail* feathers, instead of *yellow*; it *lacks* the *red band* on the *head*. It is found in the Rocky Mt. and Pacific Coast regions from British Columbia to Mexico, and east to western Texas, Nebraska, and South Dakota. In regions where the northern flicker also is found, these two species have hybridized. In the National Museum of Washington there are numerous specimens of these hybrids, where the red and black cheek-patches, the red and yellow wing-feathers and red band on the head appear in various unusual combinations.

The Flicker is a bird of distinction. A glimpse of him at once arouses interest, curiosity, and a desire for further acquaintance. He is handsome, well set up, full of vitality and power—the personification of efficiency.

We like his cheerful voice—a trifle too loud for a gentleman of refinement, but a welcome sound in the season when the whole world wishes to shout with joy at the release from winter's confinement. Thoreau wrote:

> "Ah, there is the note of the first flicker, a prolonged, monotonous *wick-wick-wick-wick-wick-wick*, etc., or, if you please, *quick, quick, quick*, heard far over and through the dry leaves. But how that single sound peoples and enriches all the woods and fields. They are no longer the same woods and fields that they were. This note really quickens 129 what was dead. It seems to put life into the withered grass and leaves and bare twigs, and henceforth the days shall not be as they have been. It is as when a family, your neighbors, return to an empty house after a long absence, and you hear the cheerful hum of voices and the laughter of children.... So the flicker makes his voice ring.... It is as good as a house-warming to all nature."[71]

We cannot repress a smile as we watch this golden-winged woodpecker striving to make a favorable impression upon Miss Flicker. He and a group of rivals take amusing, awkward attitudes, make a variety of

---

[71] From "Early Spring in Massachusetts," by H. D. Thoreau, pages 160 and 161.

noisy but pleasant calls, and without any ill-tempered quarreling, select their mates and "live happily ever after."

Though a woodpecker, the flicker departs from family habits and traditions by seeking his livelihood on the ground in preference to tree-trunks. He is a foe to the industrious ant that we were taught to admire along with the "busy bee." But ants destroy timber, infest houses, and cause the spread of aphids that are enemies of garden plants; therefore the ant's destroyer, the flicker, is a neighborhood benefactor and deserves our heartfelt protection. Professor Beal reports finding 3,000 ants in the stomach of each of two flickers and fully 5,000 in that of another.[72] These insects form almost half of this bird's food. His long, sticky tongue is especially adapted to their capture. He likes grasshoppers, crickets, beetles, and caterpillars, and while he enjoys fruit, he takes little that is of any value to man.

Most northern flickers migrate. They remain during the winter in some localities, as Cape Cod, where food is sufficiently abundant. Mr. Forbush tells of flickers that have bored holes in summer cottages on the Cape, and spent the winters in rooms which they damaged by their habit of "pecking." He states that bird-boxes containing large entrances placed on the outside of the houses or on the trees near by, would have prevented those flickers from forming the "criminal habit of breaking and entering."[73] Red-Shafted Flickers have also been found guilty of the same crime, and have entered not only dwellings, but school-houses and church steeples.[74]

Though rather shy birds, they often approach inhabited houses and frequently cause amusing situations because of their regular drumming on roof or wall. In Florida, a young woman whom I know was once aroused from her early morning's sleep by a flicker's knock, and drowsily responded with a "Come in." A friend and I, spending a week-end in an Ohio summer cottage that possessed no alarm-clock, asked to be called in time for a very early boat. We heard a knocking, arose, dressed quietly to avoid disturbing the household, and then found that our summons had

---

[72] Farmers' Bulletin 630, U. S. Dept. of Agriculture, Biological Survey.
[73] From "Useful Birds and Their Protection," by E. H. Forbush, pages 261 and 262.
[74] Farmers' Bulletin 513, U. S. Department of Agriculture, Biological Survey, H. W. Henshaw.

come from flickers on the roof, and that we had lost about two hours of precious morning's sleep.

Flickers have more local names than almost any other bird. Over one hundred names have been recorded, of which "Yellowhammer," and "Golden-winged Woodpecker," are perhaps most common.

Figure 23. Red-headed woodpecker.

The Red-Headed Woodpecker

Woodpecker Family—Picidæ

Length: About 9¾ inches; nearly as large as a robin.

General Appearance: A black and white bird *with entire head and neck bright red.*

Male and Female: Head, neck, throat, and upper part of the breast brilliant red; upper part of back and wings black; longer wing-feathers or primaries also black; lower back and secondary wing-feathers white; under parts white; tail pointed, black, margined with white. In flight, the areas of red, black, and white are very distinct.

Young: Brown heads and necks, mottled with black; upper parts of backs barred with light brown. The other parts of their bodies resemble those of their parents.

Note: No song, but a loud, cheerful *Quir-r-r-k? Quir-r-r-k?* and a drumming sound, similar to that made by other woodpeckers.

Habitat: Open woods, groves of beeches preferred.

Nest: In hollow tree-trunks or telegraph-poles.

Range: From southeastern British Columbia, to Ontario, south to the Gulf Coast, and from central Montana, Colorado, and Texas east to the valleys of the Hudson and Delaware; rare in New England. Irregularly migratory in the northern parts of its range.

This conspicuous bird is one of the handsomest members of the Woodpecker family. He is the only one really entitled to the name of Red-Headed Woodpecker. His male relatives wear only small skull-caps placed on their crowns at various angles; he possesses 132 a sort of toboggan-cap pulled down over his head and tucked into his black coat and white vest-front.

Many stories and legends are told of this woodpecker. He is the delight of children in localities where he is to be found. I remember how I used to look for the red hood and the black shawl worn over a white dress, especially noticeable in flight. I never tired of watching one of these birds approach his nest in a tall dead tree with food in his mouth. At a signal from him, his wife's red head would appear in the doorway. She would emerge; he would then enter and remain with the children until her return.

Redheads have not been popular with farmers, who have accused them of various crimes. They have been caught eating small fruit and corn on the ear, destroying both the eggs and young of other birds, and boring holes in telegraph-poles in which to build their nests. While individuals may be guilty of such misdemeanors, the redheads are probably neither so black nor so gory, except in plumage, as they are painted.

These woodpeckers are not such persistent destroyers of insects as others of their family. They have a decided preference for beetles, but eat fewer ants and larvæ than do the Downy and Hairy woodpeckers. They are exceptionally fond of vegetable food; their preference for beechnuts is very great. Dr. C. Hart Merriam states that in northern New York, where the redhead is one of the commonest woodpeckers, it subsists almost exclusively on beechnuts during the fall and winter, even pecking the green nuts before they are ripe and while the trees are still covered with leaves. He has shown that these woodpeckers invariably remain throughout the winter after 133 good nut-yields and migrate whenever the nut-crop fails.[75]

"In central Indiana during a good beechnut year, from the time the nuts began to ripen, the redheads were almost constantly on the wing; passing from the beeches to some place of deposit. They hid the nuts in almost every conceivable situation. Many were placed in cavities in partly decayed trees; and the felling of an old beech was certain to provide a feast for the children. Large handfuls were taken from a single knot hole. They were often found under a patch of raised bark, and single nuts were driven into cracks in the bark. Others were thrust into cracks in gate-posts; and a favorite place of deposit was behind long slivers on fence-posts. In a few cases grains of corn were mixed with beechnuts. Nuts were often driven into cracks in the end of railroad ties, and the birds were often seen on the roofs of houses pounding nuts into crevices between the shingles. In several instances the space formed by a board springing away from a fence

[75] Bulletin No. 37, U. S. Dept. of Agriculture, Biological Survey.

was nearly filled with nuts, and afterwards pieces of bark and wood were brought and driven over the nuts as if to hide them from poachers."[76]

In summer, Dr. Merriam has seen the redheads "make frequent sallies into the air after passing insects, which were almost invariably secured." He has also seen them catch grasshoppers on the ground in a pasture.

They are cheerful, active birds, with a call like that of a giant tree-toad. Their brilliant plumage has unfortunately made them a good target for sportsmen.

THE RED-BELLIED WOODPECKER

Woodpecker Family—Picidæ

Length: About 9½ inches.

Male: *Crown of head and back of neck bright red*, resembling slightly that of the red-headed woodpecker, but *throat and cheeks gray*; back and wings barred with white, the barring reminding one of the flicker. *Under parts gray mashed with red*; tail black and white; upper tail-coverts white, streaked with black.

Female: *Crown gray, nostrils and neck bright red.*

Notes: Mr. Frank Chapman writes of this woodpecker: "It ascends a tree in a curious, jerky fashion, accompanying each upward move by a hoarse *chu-chu*. It also utters *k-r-r-r-ring* roll and, when mating, a *whicker* call like that of the Flicker."[77]

Habitat: Open woods of deciduous trees and conifers; also groves of live-oak, palmettoes, and other southern trees, where these birds may be seen in company with flickers.

Range: From southern Canada and eastern United States southward; abundant in the Southern States; rare in New England; is found in western

[76] The Auk, IV, 194, 195, 1887. O. P. Hay.

[77] From "Birds of Eastern North America," by Frank M. Chapman, used with permission of D. Appleton & Co.

New York and south-western Pennsylvania, and Delaware, south to central Texas and the Gulf States.

Figure 24. Yellow-bellied sapsucker red-bellied woodpecker.

Professor Beal made the following report regarding this woodpecker:

"The red-bellied woodpecker ranges over the eastern United States as far west as central Texas and eastern Colorado and as far north as New York, southern Ontario, Michigan, and southern Minnesota. It breeds throughout this range and appears to be irregularly migratory. It appears to go north of its breeding range sometimes to spend the winter. Four stomachs, collected in November and December, were received from Canada, and in eight years' residence in central Iowa the writer found the species abundant every winter, but never saw one in the breeding season. It is rather more of a forest bird than some of the other woodpeckers, but is frequently seen in open or thinly timbered country. In the northern part of its range it appears to prefer deciduous growth, but in the South is very common in pine forests.

"Ants are a fairly constant article of diet. The most are taken during the warmer months. Evidently this bird does not dig all the ants which it eats from decaying wood, like the downy woodpecker, but, like the flickers, collects them from the ground and the bark of trees.

"In Florida, the bird has been observed to eat oranges to an injurious extent. It attacks the over-ripe fruit and pecks holes in it and sometimes completely devours it. The fruit selected is that which is dead ripe or partly decayed, so it is not often that the damage is serious. The bird sometimes attacks the trunks of the orange trees as well as others and does some harm. The contents of the stomachs, however, show that wild fruits are preferred, and probably only when these have been replaced by cultivated varieties is any mischief done."[78]

THE YELLOW-BELLIED SAPSUCKER

Woodpecker Family—Picidæ

Length: About 8½ inches, larger than the Downy, and smaller than the Red-headed woodpecker.

General Appearance: A medium-sized bird, with *bars*, *stripes*, and *patches* of black and white. The *scarlet crown*, the *black band across* the *breast*, and the *scarlet throat* of the males are distinguishing marks.

Male: Crown and throat bright red; bill long; head with broad black and white stripes, extending to neck. The black stripe beginning at bill unites with a *black crescent that encloses red throat*. *Breast* and *belly light yellow*; sides gray, streaked with black; back black, barred with white;

[78] Farmers' Bulletin 506, U. S. Dept. of Agriculture, Biological Survey.

wings black, with *large white patches*, white bars, and spots; *middle* tail-feathers, white and black; outer tail-feathers mostly black.

Female: Resembles male, but throat is usually white instead of scarlet.

Young: Similar to parents, but with dull blackish crowns, whitish throats, and brownish-gray breasts.

Notes: A faint call-note; a ringing call, consisting of several similar notes.

Habitat: Tree-trunks, into which these birds drill holes and thus kill the trees.

Range: Eastern North America. Breeds from the tree-belt of Canada to northern Missouri, Indiana, Ohio, mountains of Massachusetts and North Carolina; winters from Pennsylvania and Ohio Valley to the Gulf Coast, Bahamas, Cuba, and Costa Rica.

The Yellow-bellied Sapsucker is the renegade of the woodpecker family—the transgressor that has 137 called down anathemas upon all his tribe. He does more damage in some localities than others. Mr. Forbush reports that while the sapsucker has undoubtedly killed trees in northern New England where he breeds, yet in thirty years he has done no appreciable harm in Massachusetts.

Dr. Henry Henshaw, formerly Chief of the Biological Survey, writes:

> "The Yellow-bellied Sapsucker, unlike other woodpeckers, does comparatively little good and much harm." Mr. Henshaw reports 250 kinds of trees known to have been attacked by sapsuckers and left with "girdles of holes" or "blemishes known as bird-pecks, especially numerous in hickory, oak, cypress, and yellow poplar."[79]

The experience of Dr. Sylvester Judd at Marshall Hall, Maryland, was as follows:

> "In the summer of 1895 there was on the Bryan farm a little orchard of nine apple trees, about twelve years old, that appeared perfectly healthy. In the fall sapsuckers tapped them in many places, and during

[79] Farmers' Bulletin 513, U. S. Dept. of Agriculture, Biological Survey.

spring and fall of the next four years they resorted to them regularly for supplies of sap. Observations were made (October 15, 1896) of two sapsuckers in adjoining trees of the orchard. From a point twenty feet distant they were watched for three hours with powerful glasses to see whether they fed to any considerable extent on ants or other insects that were running over the tree-trunks. In that time one bird seized an ant and the other snapped at some flying insect. One drank sap from the holes thirty and the other forty-one times. Later in the day, one drilled two new holes and the other five. The holes were made in more or less regular rings about the trunk, one ring close above another, for a distance of 138 six to eight inches. The drills were about a quarter of an inch deep, and penetrated the bark and the outer part of the wood.

"In November, 1900, seven of the nine trees were dead and the others were dying. The loss of sap must have been an exhausting drain, but it was not the sole cause of death. Beetles of the flat-headed apple-borer, attracted by the exuding sap, had oviposited in the holes, and the next generation, having thus gained an entrance, had finished the deadly work begun by the sapsuckers."[80]

Mr. W. L. McAtee, of the Biological Survey, made the following report on sapsuckers:

"These birds have short, brushy tongues not adapted to the capture of insects, while the other woodpeckers have tongues with barbed tips which can be extended to spear luckless borers or other insects whose burrows in the wood have been reached by their powerful beaks. The sapsuckers practically do not feed on wood-borers or other forest enemies. Their chief insect food is ants. About 15 per cent of their diet consists of cambium and the inner bark of trees, and they drink a great deal of sap.

"The parts of the tree injured by sapsuckers are those that carry the rich sap which nourishes the growing wood and bark. Sapsucker pecking disfigures ornamental trees, giving rise to pitch streams, gummy excrescences, and deformities of the trunks. Small fruit trees, especially the apple, are often killed, and whole young orchards have been destroyed.

[80] "Birds of a Maryland Farm," by Sylvester D. Judd—Bulletin 17, Biological Survey.

> "These birds inflict much greater financial loss by producing defects in the wood of the far larger number of 139 trees which they work upon but do not kill. Blemishes frequently render the trees unfit for anything except coarse construction and fuel.
>
> "Hickory trees are favorites of sapsuckers. It is estimated that about 10 per cent of the merchantable material is left in the woods on account of bird pecks. On this basis the annual loss on hickory is about $600,000. To this must be added the loss on timber by the manufacturer."[81]

It is no wonder that war has been declared upon sapsuckers; but it is very sad that because of a lack of careful observation of the distinctive markings of tree-trunk birds, many useful woodpeckers, especially the Downy and Hairy, have been sacrificed.

Downy and Hairy Woodpeckers may be readily identified by a *broad white stripe* extending down the *center* of the *back*, a *small patch* of *red* on the *back* of the head, *pure white throats* and *breasts*, and wings *barred* with white. A *red forehead* and *crown* (and red throat of males), a *black crescent across* the *breast*, *large white patches* on the *wings*, a *back* with *black* and *white bars instead of a white streak*, differentiate this sapsucker from the Downy and Hairy woodpeckers. The yellow belly is not a conspicuous "field-mark."

There are several species of sapsucker in the West. The yellow-bellied is found in western texas; the red-naped sapsucker in the Rocky Mt. region, from British Columbia to northwestern Mexico, and from Colorado and Montana to the Cascade and Sierra Nevada Mts.; the red-breasted sapsucker in the Canadian forests of the *Pacific Coast region*, from Alaska to Lower California, east 140 to the Cascades and Sierra Nevadas; and the williamson sapsucker, from the eastern slopes of the Rocky Mts. westward to the Pacific, and from Arizona and New Mexico to British Columbia.[82] The last-named species is a great devourer of ants.

[81] Farmers' Bulletin 506, U. S. Dept. of Agriculture, Biological Survey.

[82] Bulletin No. 37, U. S. Dept. of Agriculture, Biological Survey.

THE MOURNING DOVE

Pigeon Family—Columbidæ

Length: Nearly 12 inches; tail 5½ inches.

General Appearance: A large, plump, grayish-brown bird, with a small head, a *black mark below the ear*, and a *long pointed tail*, in contrast to the round, fan-shaped tail of tame pigeons.

Figure 25. Mourning dove.

Male: Upper parts a soft grayish-brown, except the head, which is bluish-gray on the crown, with a pinkish-buff forehead, and the wings, which have long, gray primaries. *Sides of neck beautifully iridescent*, with a *small black spot below* the *ear*, an identification-mark; black spots on the lower part of breast and wings; breast with a pinkish tinge, and underneath the tail pale yellow; tail long and sharply pointed when the bird is at rest. In flight, it resembles the jay's in shape; the middle feathers are brown, like the back; outer feathers largely white; others brown, tipped with white and banded with black; feet and legs red.

Female: Duller than male, with less iridescence on neck.

Note: A soft, monotonous *coo-oo-a-coo-o-o*, uttered mournfully and with great tenderness. The sound is pleasing to some people, but unendurable to others.

Habitat: Open woodlands, or fields bordered with trees.

Range: North America. Breeds chiefly from southern Canada throughout the United States and Mexico; winters from southern Oregon, Colorado, the Ohio Valley, and North Carolina to Panama; casual in winter in the Middle States.

Mourning doves, whose "billing and cooing" have become proverbial, are as devoted pairs of lovers as may be found in the bird-world. The ardent male appears to seek the society of none except his loving mate. She seems perfectly satisfied with his attentions and evidently gives him her whole heart.

Madame Dove is a very inefficient housekeeper. Her nest, built of rough sticks, and notoriously ill-constructed—is a sort of platform on which two white eggs are laid. It is a wonder that they remain in safety long enough to be hatched, for the nests are often not more than ten feet from the ground. Were not her twin-babies as phlegmatic as their parents, they might roll out of bed and come to an untimely end.

It is fortunate that the easy-going mother does not need to prepare the bountiful repasts her family demand. She and her husband select a home-site near fields where weeds abound and where grain is raised.

The family gorge themselves upon seeds until they almost burst. Mr. Charles Nash says that "these birds are often so full of seeds that, if a bird is shot, the crop bursts open when it strikes the ground."[83]

They are of enormous economic value. Their food is almost entirely vegetable, and consists largely of the seeds of weeds that a farmer must pay to have destroyed or work hard to eradicate. Doves frequent fields of wheat, corn, buckwheat, rye, oats, and barley, but the grain they destroy is only a third of their food, and consists largely of waste kernels, according to the reports of the Department of Agriculture.[84] They like many varieties of infinitesimal 143 seeds that are eschewed by other birds; as many as 9200 seeds have been found in the stomach of one dove.

These birds have an unerring instinct for fresh water. With a peculiar, whistling sound, they fly at nightfall to a spring or pool for a cool drink before retiring. Hunters are said to have watched them and thus found springs for their needs.[85]

Doves eat quantities of gravel to aid in the digestion of their epicurean feasts. They are fond of dust-baths. They also indulge in queer, senseless-looking acrobatic performances, which appear like attempts at gymnastics.

THE BELTED KINGFISHER

Kingfisher Family—Alcedinidæ

Length: About 13 inches—a rather large, stocky bird.

General Appearance: A large bluish-gray and white bird, with a *very large crested head*, a *long bill*, and a short tail.

Male: Bluish-gray above, becoming darker on the wings; a ragged-looking crest on an unusually large head; a white spot in front of each large dark eye; small flecks on the wings; tail bluish-gray, flecked and barred with white; *throat white*, a *band* of *white extending nearly around the*

[83] "Useful Birds and Their Protection," E. H. Forbush, page 324.
[84] Farmers' Bulletin 513, U. S. Dept. of Agriculture, Biological Survey.
[85] "Life Histories of North American Birds,"—Maj. Chas. Bendire.

neck; a *broad band* of *bluish-gray extending across the breast*; under parts white, except the sides, which are bluish-gray; feet relatively small, but with long, strong nails.

Female: Similar to the male, except for a *band* of *reddish-brown across the breast*, extending to the sides, and forming a fourth belt; a white belt at the throat, then gray, white, and reddish-brown belts. Unlike most birds, the female kingfisher is more highly colored than the male.

Note: A long harsh rattle, similar to the sound made by two bones or smooth sticks in the hands of a boy, or to the noise of a policeman's rattle.

Habitat:

> "By a wooded stream or a clear cool pond,
> Or the shores of a shining lake."

Range: North America, and northern South America. Breeds from Alaska and northern Canada to the southern border of the United States; winters from British Columbia, central United States to the West Indies, Colombia, and Guiana, irregularly to Massachusetts, New Hampshire, and Ontario.

This self-appointed guardian of our streams and lakes is clad in a suit of gendarme blue. He wears a sharp two-edged sword in his cap, and carries a rattle in his throat.

He is a perfect example of "Watchful Waiting," as he sits motionless on a bough overhanging a stream, with his fierce eyes fixed intently upon the waters beneath him. When an unwary fish swims by, this blue-coat plunges after it and spears it with deadly accuracy. If small, the fish is swallowed whole; if large, it is beaten to death against a tree, and devoured with difficulty. When fish are not obtainable, the kingfisher will eat frogs and crustaceans, and sometimes grasshoppers, crickets, and beetles. Fish, however, are his favorite food.[86]

---

[86] Educational Leaflet No. 19, National Association of Audubon Societies.

Figure 26. Kingfisher.

The nest is as unusual and interesting as the bird himself. It consists of a tunnel excavated in a bank by the long knife-shaped bills of the kingfisher and his mate. A cavity of good size must be hollowed out to accommodate so large a bird and a family of from five to eight lusty youngsters. They are lively and quarrelsome; they set up a great clamor when Father or Mother arrives with an already-prepared fish-dinner. Dr.

Francis H. Herrick, in his delightful book, "The Home Life of Wild Birds," tells of his observations of a kingfisher's nest and nesting habits as follows:

> "The nest had a 4 inch bore; 4 feet from the opening was a vaulted chamber 6 inches high and 10 inches across...."

A series of rattles announced the approach of the parent bird "who came at full tilt with a fish in her bill, making the earth resound." In response came "muffled rattles of 146 five young kingfishers, who issued from their subterranean abode.... With a rattle in shrillest crescendo, she bolted right into the hole, delivered the fish, remained for half a minute, then came out backwards, turning in the air as she dropped from the entrance, and with a parting rattle was off to the river."

There were five babies in what Dr. Herrick called the "King Row." They were amusing to look at as they sat back on their legs; the bill of one nestling protruded above the shoulder of the bird in front of it. They never seized their food (fish) of their own accord.

> "It was necessary to open their bills and press the food well down into the distensible throats." Raw meat was rejected, but they throve on fish. "Kingfishers' throats are lined with inwardly projecting papillae so that when a fish is once taken in its throat, it is impossible for it to escape."[87]

The young kingfishers that Dr. Herrick observed became very tame. He is pictured with them on his hand, his shoulder, and on both knees.

While kingfishers do less good than most of our feathered benefactors, they do not destroy enough fish to be a detriment to the fishing interests of lakes and streams. They are true sportsmen, whose presence we should miss when we followed the rod and creel. We are forced to respect their prowess, and we may apostrophize them in the words of Izaac Walton:

---

[87] From "The Home Life of Wild Birds," by Francis H. Herrick. Used with the permission of the author, and of his publishers, G. P. Putnam & Co.

> "Angling is an Art, and you know that Art better than others; and that this is the truth is demonstrated by the fruits of that pleasant labor which you enjoy."

The Field Sparrow

Finch Family—Fringillidæ

Length: About 5½ inches.

General Appearance: A small brown bird with a reddish back and bill, and a buff breast without spots or streaks.

Male and Female: Top of head reddish-brown; sides of head, nape of neck, and line over eye gray; bill reddish-brown; back reddish-brown, streaked with black and gray; rump brownish-gray; wings and tail brown, some wing-feathers edged with gray; sides and breast washed with buff.

Song: A sweet trill, consisting of the syllable dee repeated a number of times. It varies with different individuals, but is phrased somewhat as follows: Dee′-dee′-dee′, de′-de, de′-de, de′-de, de′-de, de′-de, de′-de.

Habitat: Old overgrown pastures containing clumps of bushes, preferred to cultivated fields. This sparrow is not accurately named, for it is not strictly a bird of the fields.

Range: Eastern North America. Breeds from southern Minnesota, Michigan, Quebec, and Maine to central Texas, Louisiana, and northern Florida; winters from Missouri, Illinois, southern Pennsylvania, and New Jersey to the Gulf Coast.

Some gorgeous but noisy birds, like blue jays, peacocks, and parrots, please only the eye; many quietly-dressed but sweet-voiced songsters are a delight to the ear. To the latter class belongs the Field Sparrow, a gentle little bird, so rarely seen as to recall to our minds the lines:

> "Shall I call thee Bird
> Or but a wandering Voice?

Even yet thou art to me
No bird, but an invisible thing,
A voice, a mystery."

Figure 27. Field sparrow.

It was several years after I had learned to love the sweet, tender song of the field sparrow that I had my first glimpse of the singer. He is a very real and delightful part of our April meadows, where he lives his serene life.

The Vesper Sparrow

Finch Family—Fringillidæ

Length: A little over 6 inches; slightly larger than the field sparrow.

Male and Female: Brownish-gray above, with faint streaks of black and buff; wings brownish, with *bright reddish-brown shoulders*, giving this sparrow the name of *Bay-Winged Bunting*. Under parts white, the sides and breast streaked with black and buff; tail brownish, with *outer tail-feathers mostly white, and conspicuous in flight*.

Song: A plaintive minor strain, usually consisting of two notes followed by a trill. The syllables sound like *Sweet'-heart, I love you-you-you-you-you*.

Habitat: Grassy pastures and plowed fields, usually in the open, away from farmhouses and out-buildings.

Range: Eastern North America. Breeds from central Canada south to eastern Nebraska, central Missouri, Kentucky, Virginia, and North Carolina, west to western Minnesota; winters from the southern part of its breeding range to the Gulf Coast, west to central Texas.

The Vesper Sparrow is very easy to identify because of its white tail-feathers. They show conspicuously as the bird flutters beside hedges that border fields, frequently keeping just ahead of the observer.

The bird is less attractive in appearance than the other familiar sparrows, but has to my mind the sweetest voice of all the sparrows that I know except the fox sparrow. Its song is pensive and tender, with a spiritual quality 150 that gives it a high rank. The song sparrow's lay usually consists of three similar notes sung in a major key with a rising inflection, and followed by a cheerful trill; the vesper sparrow's song generally has two plaintive notes preceding a trill, sung in a minor key. It is particularly beautiful and uplifting when several vesper sparrows are singing at sunset.

Figure 28. Vesper sparrow.

The Vesper Sparrow

When the meadows are brown or flushed with green
And the lark's glad note rings clear,—
When the field sparrow's voice like a silver bell
Chimes a melody sweet to hear,—

A small brown bird with bay-capped wings
And feathers white in his tail,
Flutters along by a roadside hedge
And alights on a zigzag rail,
And breathes forth a song entrancing,
Of a beauty surpassed by few—
A wistful, plaintive, minor strain—
"O Sweetheart, I love you!"
When a mist of green o'erspreads the trees,
And corals and rubies gay
Are hung on the maple and red-bud boughs,
And the brooks are babbling away,—
When the setting sun goes down in a glow
Of the purest primrose gold,
And the pearly east reflects a flush
From the glories the west doth hold,—
This brown bird then, with a soul in his voice,
Sings to his mate so true
The tenderest song of the April choir—
"O Sweetheart, I love you!"

THE CHIPPING SPARROW

Finch Family—Fringillidæ

Length: A little over 5 inches; the smallest of our common sparrows.

Male and Female: *Crown reddish-brown, bill black; a black line extending through the eye*; a *gray line above the eye*; back, wings, and tail brown; tail forked; rump gray; *breast pale gray without streaks or spots*. In the fall, the reddish crown becomes brown, streaked with black.

Call-note: *Chip-chip.*

Song: A monotonous trill, *Chippy-chippy-chippy-chippy-chippy-chippy-chippy*, more like the metallic sound made by a locust than the song of a bird.

Habitat: A "doorstep" bird that loves to spend the spring and summer near man. It is found in gardens, orchards, and plowed fields.

Nest: An unusually dainty nest made of grass and fine root-fibers, lined with horsehair, which has given to the chipping sparrow the name of "hair-bird." The nest is built in trees or low bushes, sometimes very near the ground.

Eggs: Four or five pale-green eggs, mottled with dark markings.

Range: North America, from central Canada to Central America; commonest in the east.

Figure 29. Chipping sparrow.

This gentle, trustful sparrow is a general favorite. He is an unobtrusive little bird, seemingly contented to occupy his place in the world near to the haunts of man, unconsciously doing his important work without 152 noisy demonstration. Like the brown creeper and the phœbe, he is of great economic value; like them, he is not particularly interesting, and he is without skill as a songster. But his monotonous trill is a pleasant part of the spring chorus, and his presence in our yards we should sorely miss.

Mr. Forbush speaks in high praise of this bird's usefulness. He claims that the chippy is "the most destructive of all birds to the injurious pea-louse, which caused a loss of three million dollars to the pea-crop of a single state in one year."[88] This sparrow eats the grubs that feed on beet-leaves, cabbages, and other vegetables; he devours cankerworms and currant worms, besides gypsy, brown-tail, and tent caterpillars, any one of which would entitle him to our protection. In the fall, with the decrease of life in the garden, he takes to the fields, where like other sparrows he feasts on seeds.

If it were more generally known how invaluable chipping sparrows are, people would guard them more carefully from marauding cats. I wish it might become as unlawful to let cats stalk abroad during the nesting season as it is to allow unmuzzled dogs to go about freely during dog-days. I know of a bird-lover near Painesville, Ohio, who never during nesting-time allowed her pet cat to stir outside of a good-sized enclosure without a weight attached to his collar. Some people have put bells on their cats' necks, but while that is efficacious in alarming parent-birds, it is of no value in preventing the slaughter of young birds that have just left the nest. Mr. Forbush has written an appeal, which I wish was more widely known and heeded. It is called "The Domestic Cat" and 153 was published under the direction of the Massachusetts State Board of Agriculture.

Mr. Forbush wrote to such eminent experts and authorities on bird-life as Robert Ridgway, Dr. Frank M. Chapman, Dr. Witmer Stone, Dr. Henry W. Henshaw, Dr. William T. Hornaday, John Burroughs, William Dutcher, T. Gilbert Pearson, Dr. George W. Field, Dr. C. F. Hodge, Ernest Harold

---

[88] From "Useful Birds and their Protection," by E. H. Forbush.

Baynes, Mrs. Mabel Osgood Wright, and others, for their opinions regarding the relative destructiveness of cats to the bird-life of the country. They were unanimous in their denunciation of cats as the "greatest destructive agency to our smaller song and insectivorous birds."

Mrs. Wright says: "If the people of the country insist upon keeping cats in the same number as at present, all the splendid work of Federal and State legislation, all the labors of game- and song-bird protective associations, all the loving care of individuals in watching and feeding, will not be able to save our birds in many localities."

Young chipping sparrows are spoiled bird-babies. They "tag" their gentle little parents about with unusual persistence, knowing that they will get what they demand. They frequently look as if they might not turn out to be excellent bird-citizens like their ancestors. When a noted ornithologist first saw Mr. Horsfall's original drawing of the accompanying family of chipping sparrows he remarked, "That baby looks a million years old and steeped in sin!" But the duties of parenthood sober the youngsters, and the following year, they become in turn pleasant, docile, lovable little "Bird Neighbors."

The White-Throated Sparrow

Finch Family—Fringillidæ

Length: About 6¾ inches.

General Appearance: One of the larger sparrows, with a *black and white striped crown*, a *white throat*, and a *yellow spot before the eye*.

Male and Female: Striped crown, with a *narrow white line* in the *center*, a broad black stripe on each side of the white; a broad white stripe *over* the *eye* edged with a narrow black line; *a yellow spot in front of the eye*, and at the outer curve of the wing. Back brown, streaked with black; rump and tail grayish-brown; wings with two white bars; breast gray, becoming whitish on the belly; sides brownish.

Notes: A sharp *chip* for the alarm-note; low, pleasant twitterings.

Song: A sweet whistle, usually pitched high. It consists of two or three notes that vary considerably. Sometimes the first note is an octave below the second; at other times it is a few tones higher than the second. I heard one recently that sang a perfect monotone as follows: *Dee, dee, de'-de-de, de'-de-de, de'-de-de*. The song has been interpreted in Massachusetts as

Sam, Peabody, Peabody, Peabody
/
Old /
and the bird is known as the "Peabody Bird."

Habitat: Hedgerows and thickets along roadsides, in parks, on estates, and in woods.

Range: Eastern and central North America. Breeds from north-central Canada to southern Montana, central 155 Minnesota and Wisconsin, and mountains of northern Pennsylvania, New York, and Massachusetts; winters from Missouri, the Ohio Valley, southern Pennsylvania, Massachusetts, (casually in Maine), south to northeastern Mexico and Florida.

THE WHITE-CROWNED SPARROW

Finch Family—Fringillidæ

Length: Nearly 7 inches; a little larger than the white-throated sparrow.

Male and Female: Crown *white*, bordered on each side by a broad black stripe that extends from bill in front of the eye; a broad white stripe borders each black stripe; a narrow line of black borders the white. *No yellow on head or wing* like that of the white-throated sparrow. Cheeks, neck, throat, and under parts gray; belly white, sides buff; back, wings, and tail brown; back streaked; wings with two white bars.

Song: A sweet whistled strain.

Habitat: Thickets, woods, and fields.

Range: Breeds in Canada, the mountains of New Mexico, Colorado, Wyoming, and Montana, and thence to the Pacific Coast; winters in the southern half of the United States and in northern Mexico.

The White-crowned Sparrow is considered by some admirers to be the handsomest member of the sparrow tribe. It is not widely known in the East, and is sometimes confused with the white-throat. The gray throat of the white-crown and the absence of yellow on the wing and near the eye, distinguish it from the white-throat.

In Bulletin 513 of the Biological Survey occurs this description of the white-crown: "This beautiful sparrow is much more numerous in the western than in the eastern States, where indeed it is rather rare. In the East it is 157 shy and retiring, but it is much bolder and more conspicuous in the far West and often frequents gardens and parks. Like most of its family it is a seed-eater by preference, and insects comprise very little more than 7 per cent of its diet. Caterpillars are the largest item, with some beetles, a few ants and wasps, and some bugs, among which are black olive scales. The great bulk of food, however, consists of weed seeds, which amount to 74 per cent of the whole. In California this bird is accused of eating the buds and blossoms of fruit trees, but buds or blossoms were found in only 30 out of 516 stomachs, and probably it is only under exceptional circumstances that it does any damage in this way. Evidently neither the farmer nor the fruit-grower has much to fear from the white-crowned sparrow. The little fruit it eats is mostly wild, and the grain eaten is waste."

THE PURPLE FINCH

Finch Family—Fringillidæ

Length: About 6¼ inches; a little smaller than the English sparrow.

Adult Male: Body largely raspberry- or rose-red, streaked with brown. For two seasons the male is a brown sparrowlike bird, with a yellowish-olive chin and rump; the third season his body seems to have been washed with a beautiful *red, not purple*, the color richest on his head, breast, and rump. Head slightly crested; bill thick, with bristles at nostrils; cheeks and back brownish; under parts grayish-white; wings and tail brownish, edged with red; tail forked.

Female: Decidedly sparrowlike; body grayish-brown, heavily streaked, lighter underneath; patch of light gray extending from eye, another from beak; wings dark grayish-brown, with indistinct gray bands. She is not unlike the song sparrow, except for the absence of the three black spots on breast and throat.

Call-note: A sharp, metallic *chip*.

Song: A clear, sweet, joyous warble.

Habitat: Woods, orchards, and gardens.

Range: Eastern North America. Breeds in central and southern Canada, and northern United States, in North Dakota, central Minnesota, northern Illinois, and New Jersey, Maine, Massachusetts, the Pennsylvania mountains, and Long Island; winters from considerably north of the southern boundary of its breeding-range to the Gulf Coast, from Texas to Florida.

None of our smaller finches, except the goldfinch and indigo bunting are more beautiful in color than the purple finch which wears a Tyrian purple, rather than the shade we commonly know.

Few members of the family sing more sweetly and joyously than this songster of the treetops. His delightful warble resembles somewhat the song of the rose-breasted grosbeak, and attracts attention wherever the bird is to be found. Several purple finches singing from neighboring elm trees at once, makes a May or June concert not easily excelled. Mr. Forbush says: "The song of the male is a sudden, joyous burst of melody, vigorous, but clear and pure, which no mere words can do justice. When, filled with ecstasy, he mounts in air and hangs with fluttering wings above the trees where sits the one who holds his affections, his efforts far transcend his ordinary tones, and a continuous melody flows forth, until, exhausted with his vocal efforts, he sinks to the level of his spouse in the treetop. This is a musical species, for some females sing, though not so well as the males."[89]

[89] From "Useful Birds and Their Protection," by E. H. Forbush.

The bird has been accused of eating the buds of fruit and shade trees, especially elms, and while he is at times guilty, he is not condemned by those who know his food-habits best, but commended for his fondness for weed seeds, especially ragweed, and for destroying plant-lice, cankerworms, cutworms, and ground beetles.[89]

His cousin, the house finch, or linnet of California, who is brighter in color, is more beloved by tourists and more hated by fruit-growers than almost any bird in the state. Professor Beal writes: "This bird, like the other 160 members of its family, is by nature a seed-eater, and before the beginning of fruit-growing in California probably subsisted upon the seeds of weeds, with an occasional wild berry. Now, however, when orchards have extended throughout the length and breadth of the state and every month from May to December sees some ripening fruit, the linnets take their share. As their name is legion, the sum total of the fruit that they destroy is more than the fruit-raiser can well spare. As the bird has a stout beak, it has no difficulty in breaking the skin of the hardest fruit and feasting upon the pulp, thereby spoiling the fruit and giving weaker-billed birds a chance to sample and acquire a taste for what they might not otherwise have molested. Complaints against this bird have been many and loud.... Whatever the linnet's sins may be, grain-eating is not one of them. In view of the great complaint made against their fruit-eating habit, the small quantity found in the stomachs taken is somewhat of a surprise. When a bird takes a single peck from a cherry or an apricot, it spoils the whole fruit, and in this way may ruin half a dozen in taking a single meal. That the damage is often serious no one will deny. It is noticeable, however, that the earliest varieties are the ones most affected; also, that in large orchards the damage is not perceptible, while in small plantations the whole crop is frequently destroyed."[90]

In spite of this troublesome habit, the linnet is a most engaging little bird. Its sweet bubbling song, not unlike that of the purple finch, adds much to the charm of California.

---

[90] Farmers' Bulletin 630, U. S. Dept. of Agriculture, Biological Survey.

The Towhee or Chewink Called Also Ground Robin and Charee

Finch Family—Fringillidæ

Length: About 8½ inches; smaller than the robin and larger than the oriole.

General Appearance: A black bird with reddish-brown sides, black breast, and white belly; outer tail-feathers tipped with white.

Male: Head, back, throat, and breast, a glossy black; wings black, outer feathers edged with white; tail black, outer edge of outer feather white; three other feathers partly white, decreasing in size toward middle of tail; belly white; eyes dark red.

Female: Brownish, where male is black. The young are streaked with black.

Call-note: A cheerful cha-ree, uttered with a rising inflection. The note is also interpreted as tow hee′? chewink′? jaree′? An engaging trait of this bird is his almost invariable response to one imitating his note.

Song: Two notes, followed by a trill. The song may be translated into chip-chur, pussy-pussy-willow.

Habitat: Woodlands, where he is first found in April scratching among old leaves like fox sparrows, white-throats, and other members of his family.

Range: Eastern North America. Breeds from southern Canada and Maine to central Kansas and northern Georgia; winters from southeastern Nebraska, the Ohio and Potomac Valleys to central Texas, the Gulf Coast, and southern Florida.

The white-eyed towhee is found on the Atlantic Coast region from about Charleston, South Carolina, 162 to southern Florida. He resembles his northern cousin except that his eyes are white, and that his wings and tail have less white on them. There are several species of towhee in our western states.

Figure 30. Towhee.

Before the trees are in leaf, there appears in our April woods a lively, trim, and attractive bird who makes himself known in no uncertain manner. So bustling and energetic is he, so cheerful and self-confident, without unpleasant aggressiveness, that he always attracts attention. The uninitiated frequently call him an oriole, whom he does resemble in having a glossy black head, throat, back, and tail, and white markings on his wings, with reddish-brown like that of the orchard oriole on his sides; but there the resemblance ceases, for the oriole has in addition a reddish-brown breast, belly, and rump. Then, too, the towhee arrives early, before larvæ have hatched; the oriole arrives in May, when swarms of insects have begun their work of fertilizing blossoms of fruit trees.

Professor Beal writes of the towhee as follows: "After snow has disappeared in early spring, an investigation of the rustling so often heard among the leaves near a fence or in a thicket will frequently disclose a towhee at work scratching for his dinner after the manner of a hen; and in these places and along the sunny border of woods, old leaves will be found overturned where the bird has been searching for hibernating beetles and larvæ. The good which the towhee does in this way can hardly be overestimated, since the death of a single insect at this time, before it has had an opportunity to deposit its egg, is equivalent 163 to the destruction of a host later in the year."[91]

While attending to business, this ground robin seems most materialistic and worldly-minded; but when satisfied with his quest for food, "a change comes over the spirit of his dreams." He perches upon a low bough; in a sweet and joyous song he reveals his passionate devotion to his mate, and brings pleasure to listeners whose ears are attuned to the sounds of Nature.

[91] Farmers' Bulletin 630, U. S. Department of Agriculture, Biological Survey.

PART 3 - DESCRIPTIONS AND BIOGRAPHIES OF OUR LATER SPRING BIRDS

Chapter 7

INTRODUCTION - LATER SPRING BIRDS

Spring comes with a rush in some parts of our country and remains but a short time, so closely does Summer follow in her footsteps. But in New England, New York, northern Pennsylvania, Ohio, and neighboring states, her approach is more gradual and restrained.

When maple and red-bud have laid aside their corals and fruit-trees have donned their robes of white and shell-pink; when the woods show again a flush of tender green, Spring arrives. She has long been heralded by early choristers; she is now accompanied by a host more wonderful than retinue of kings, so varied is their dress and so sweet their triumphal music. Grove and orchard are alive with happy-hearted birds, who help to make May the loveliest month of the year.

First come the swallows, skimming over pools and circling above meadows—embodiment of grace, gladdening the world with their joyous twitterings. Swifts, nighthawks, and whip-poor-wills make nightfall vocal. Little house wrens, each a fountain of bubbling music, take up their abode near our homes.

Cuckoos slip quietly from tree to tree; thrashers and catbirds seek thickets or perch on treetops, to sing like their celebrated cousins, the mockingbirds. Shy ovenbirds and lustrous-eyed thrushes return to live in the woods, or pass through them as they journey to their northern homes.

The advent of the tanager in his flashing scarlet, and the grosbeak with his glowing rose bring to 168 every bird-lover "a most pointed pleasure." With Stevenson he may say, [They] "stab my spirit broad awake."

Vireos and wood pewees appear in the groves; warblers flit from treetop to treetop, many of them on their way to northern woods. Orioles in the elms and orchards shout with joy; bobolinks bubble and tinkle in the meadows; indigo buntings and kingbirds greet us from roadsides, and Maryland yellow-throats from thickets. Goldfinches hold their May festival, and choose their mates as they sing with joyous abandon. The earth is fresh and beautiful, with promise of a glad fulfillment near at hand.

Figure 31. Tree swallow.

Chapter 8

DESCRIPTIONS AND BIOGRAPHIES OF LATER SPRING ARRIVALS

THE TREE SWALLOW

Swallow Family—Hirundinidæ

Length: About 6 inches.

General Appearance: Bluish-green above; pure white underneath, from beak to tail; *tail not deeply forked*; wings very long.

Male and Female: Back, a dark, glistening green, giving this swallow the name of "The Green-backed Swallow"; the snowy white under parts give it the names of "White-breasted Swallow" and "White-bellied Swallow." The green and white are about equally distributed; the green on the head resembles a close-fitting skull-cap, pulled down below the eyes. Wings, *very long and powerful* (nearly 4¾ inches), extending beyond the ends of the forked tail. Bill short, *very wide at base*. Feet small and weak—used only when resting, as swallows are generally on the wing.

Young: Brownish-gray, white beneath.

Note: A pleasant twitter.

Flight: Swift, in great circles.

Habitat: Tree swallows are seen along roadsides, and near swamps and thickets. They formerly nested in dead trees, in woodpeckers' holes, or any available hollow. They now take kindly to nesting-boxes. They have "roosts" at night where they resort in great numbers, especially on their way south in the late summer. They have a great fondness for telegraph-wires. During the fall migration, long chains of these swallows are festooned on the wires during the daytime. At night they disappear to their roosts, preferably near marshes. They are a sight to be remembered 170 in the Jersey marshes, which Mr. Horsfall's accompanying drawing depicts.

Range: North America from Alaska and northern Canada to southern California, Colorado, Kansas, Missouri, and Virginia. They winter from central California, southern Texas, southern parts of the Gulf States and southeastern North Carolina, south over Mexico, Guatemala, and Cuba; sometimes in New Jersey. They eat bayberries that grow along the coast, and thus are able to remain farther north in winter than their relatives.

First of the swallow host to speed northward is the Tree Swallow, that migrates in April, as soon as a sufficient number of insects have hatched to furnish a living for these almost wholly insectivorous birds. Their cheerful twitter and beautiful circling flight make them very welcome.

Swallows have always been regarded with favor. They were formerly considered a good omen, and were thought to bring fair weather and prosperity. I shall always remember the welcoming swallow that met our ship near the Scilly Islands one June day, and preceded us without resting for long hours as we voyaged close to the shore of England. It seemed to presage the good fortune that followed us.

Swallows fly with their broad beaks ready to open, and catch unwary insects with great ease. They rise early and continue their ceaseless quest for small beetles, flies, mosquitoes, and other insects. Professor Beal says: "Most of these are either injurious or annoying, and the numbers destroyed by swallows are not only beyond calculation 171 but almost beyond imagination."[92] He pleads for the protection of all swallows and suggests that the "white-bellied swallows" be supplied with boxes similar to those

[92] Farmers' Bulletin 630, U. S. Biological Survey, U. S. Dept. of Agriculture.

constructed for bluebirds, only placed at a greater elevation and protected from cats.

Tree swallows are the first to come and first to go. Before the summer has really arrived, as early as July first, they begin to flock and form great colonies that may be seen migrating during the daytime.

THE BARN SWALLOW

Swallow Family—Hirundinidæ

Length: About 7 inches; an inch longer than the tree swallow because of longer tail; body nearly the same size.

General Appearance: Upper parts a glossy bluish-black; under parts reddish-brown and buff; tail deeply forked.

Male: Forehead and throat bright reddish-brown; breast, belly, and feathers under wings a light brown, becoming buffy; breast and throat separated by an indistinct dark band; upper parts a shimmering bluish-black; tail very deeply forked—the proverbial "swallow-tail"; rounded white spots on the inner web of all except the middle tail-feathers.

Female: Resembles male, though paler in color; outer tail-feathers a little shorter.

Young: Backs duller, breasts paler, tail-feathers shorter than those of adult male.

Notes: A clear, sweet call, and a joyous, musical twitter—weet-weet, or twit-twit.

Flight: Long, sweeping curves that are beautiful to see. The bird shows first his blue back, then his soft brown breast. He flies nearer the ground than other swallows, and surpasses them all in his power of flight. Imagine the number of miles he travels in a day!

Habitat: Fields and farm-lands; also the vicinity of ponds or other breeding-places of insects. The nest of mud is usually fastened to a rafter of a barn. These swallows often nest in colonies.

Range: North America, from northwestern Alaska and Canada, to southern California and southwestern Texas, northern Arkansas and North Carolina. They do not breed in the southeastern part of the United States. They winter in South America.

Figure 32. Barn swallow.

Most beautiful of all the swallows is this bluebird fleet of the summer time. It is associated in my mind with shining pools rimmed with iris; with fragrant lilac-bushes, blossoming apple-trees, and waving fields of grain near farm-buildings. Its sweet voice and marvelous flight bring poetry into the prosaic life of the farm.

Burroughs characterizes the swallow delightfully in "Under the Maples." He says:

> "Is not the swallow one of the oldest and dearest of birds? Known to the poets and sages and prophets of all peoples! So infantile, so helpless and awkward upon the earth, so graceful and masterful on the wing, the child and darling of the summer air, reaping its invisible harvest in the fields of space as if it dined on sunbeams, touching no earthly food, drinking and bathing and mating on the wing, swiftly, tirelessly coursing the long day through, a thought on wings, a lyric in the shape of a bird! Only in the free fields of the summer air could it have got that steel-blue of the wings and that warm tan of the breast. Of course I refer to the barn swallow. The cliff swallow seems less a child of the sky and sun, probably because its sheen and glow are less, and its shape and motions less arrowy. More varied in color, its hues yet lack the intensity, and its flight the swiftness, of those of its brother of the hay-lofts. The tree swallows and the bank swallows are pleasing, but they are much more local and restricted in their ranges than the barn-frequenters. As a farm boy I did not know them at all, but the barn swallows the summer 174 always brought. After all, there is but one swallow; the others are particular kinds that we specify."[93]

The Purple Martin

Swallow Family—Hirundinidæ

Length: About 8 inches, the largest of the six common species of swallow. Wings nearly 6 inches long—very large when spread.

---

[93] Used with permission of the Houghton Mifflin Co., the authorized publishers.

Male: *Glossy purplish-black head, body, and shoulders; wings and tail duller. No reddish-brown or white. Tail forked.*

Figure 33. Purple martin.

Female: Bluish-black head and back; black wings and tail; brownish-gray throat, neck, and sides, mottled with white-tipped feathers; belly, grayish-white.

Young: Similar to female.

Note: A sweet, rich, joyous warble. Mr. Forbush describes it as "a full-toned chirruping carol, musical and clear, beginning *peuo-peuo-peuo*."[94]

Habitat: Farm-lands and the vicinity of dwellings shaded by trees. These birds were formerly more numerous in the North than at present. They are more abundant in the South than in the North.

Nests: Made of twigs, grass, straw, or leaves, placed in gourds or martin-houses. Martins are very social and seem to revel in large "bird-apartment-houses." They formerly nested in hollow trees or caves.

Range: North and South America, except Pacific Coast region. They breed in southern Canada, east of the Rockies; in the United States from Montana and Idaho, south to the Gulf Coast, Florida, and Mexico. They winter in Brazil. A western martin is found on the Pacific Coast.

Purple Martins have long been favorites. Mr. Dutcher tells us that Indians, keen observers of 176 nature, realized that it was beneficial to have them near their long-houses. They therefore hung hollowed gourds to entice them. Southern negroes have done likewise. They sometimes suspend a number of gourds from crossbars surmounting a pole, to form nesting-sites for a small colony.

Martins form an ideal community—busy, happy, harmonious—unless English sparrows attempt to evict them and appropriate their homes. Martin-houses and bluebird nesting-boxes seem to be the envy of these pugnacious sparrows. Martins attack crows and hawks but cannot endure the persecutions of the English sparrow.

Martins are so useful that they should be protected and encouraged whenever possible. A friend of mine told me that she was never obliged to have her trees sprayed while the martins remained. They feed on wasps, bugs, and beetles, several varieties of which are harmful, and they devour many flies and moths.

Dr. Dutcher quotes from Audubon regarding the flight of martins as follows:

> "The usual flight of this bird ... although graceful and easy, cannot be compared in swiftness with that of the Barn Swallow. Yet the martin is

---

[94] From "Useful Birds and Their Protection," by E. H. Forbush, page 348.

fully able to distance any bird not of its own genus. They are very expert at bathing and drinking while on the wing, when over a large lake or river, giving a sudden motion to the hind part of the body, as it comes in contact with the water, thus dipping themselves in it, and then rising and shaking their body, like a water spaniel, to throw off the water."[95]

The Cliff or Eave Swallow

Swallow Family—Hirundinidæ

Length: About 6 inches; one inch smaller than the barn swallow, and two inches smaller than the martin.

General Appearance: A multi-colored swallow—a sort of combination of barn swallow and martin, with areas and patches of dark blue, chestnut, gray, and white, and *bright reddish-brown upper tail-coverts*, that differentiate it from the other swallows.

Male and Female: Forehead creamy white, head bluish-black; throat and cheeks reddish-brown; a brownish ring about the neck shading to gray; back bluish-black streaked with white; breast gray with a wash of brown, and a blue-black patch where the throat joins the breast; wings and tail brownish; tail only slightly forked.

Note: A harsher, less musical note than that of the barn swallow and martin.

Habitat: Meadows and marshes. These swallows formerly nested in cliffs; now they build under eaves of buildings.

Nests: Curiously shaped pouches of mud that make one think of protuberant knot-holes, or of flasks made of skin. The nests vary with the shape of the places to which they are fastened. Eave swallows also nest in colonies.

---

[95] Educational Leaflet No. 13, of the National Association of Audubon Societies.

Range: North America. Breed from central Alaska and north-central Canada over nearly all the United States except Florida and the Rio Grande Valley. They probably winter in Brazil and Argentina.

Mr. Forbush writes about the Cliff or Eave Swallow as follows:

> "When the first explorers reached the Yellowstone and other western rivers, swallows were found breeding on the precipitous banks. As settlers gradually worked their way westward, the swallows found nesting-places under the eaves of their rough buildings. In these new breeding-places they were better protected from the elements and their enemies than on their native cliffs and so the Cliff Swallow became the Eave Swallow, and, following the settlements, rapidly increased in numbers and worked eastward."[96]

These swallows were very numerous fifty years ago. It is now generally conceded that English sparrows are largely responsible for their decrease. It is greatly to be deplored, for swallows add much to the charm of out-door life, and subtract many annoyances in the form of insect pests, especially flies and mosquitoes.

The Bank Swallow

Swallow Family—Hirundinidæ

Length: A little over 5 inches; the smallest of the six common swallows.

General Appearance: *Brownish-gray above; band of same color across breast*; throat and under parts white. The gray head and white throat form a cap similar in effect to that of the tree swallow.

Note: A twitter, less pleasing than that of the martin and the barn swallow.

Habitat: Sandy banks of rivers, and shores of lakes.

---

[96] From "Useful Birds and Their Protection," by E. H. Forbush, page 346.

Nests: In holes made in sand-banks.

Range: North and South America. Breeds from the tree-regions of Alaska and Canada to southern California, Arizona, Texas, Louisiana, and Virginia. It migrates through Mexico and Central America and probably winters in northern South America to Brazil and Peru.

The rough-winged swallow resembles the bank swallow so closely that it is difficult to distinguish them, unless one can see the darker breast and throat of the rough-wing and the *absence of a dark band across the breast*. Upon careful examination of the latter species, each long outer wing-feather is discovered to have a rough saw-tooth edge.

The habits of the birds are similar, though the rough-wings, like phœbes, nest not only in banks, but against stone walls and stone bridges. They have a more restricted range than barn swallows. They breed from southern Canada to northern Florida and southern California, and winter in Mexico and Central America.

The Chimney Swift

Swift Family—Micropodidæ

Length: About 5½ inches; wings nearly 5 inches long.

General Appearance: In the sky, the swift looks unlike any other bird. The wings are long and flap like those of a mechanical toy-bird. The tail appears *rounded*, not *forked*, like those of swallows.

Male and Female: Brownish-gray, lighter gray on throat; a black spot before each eye; wings longer than tail; tail short, with ribs of the feathers extending beyond the vanes, giving the effect of sharp needle- or pin-points. The bird has a *sooty* appearance.

Note: A noisy, incessant twitter.

Flight: Rapid, and seemingly erratic and aimless. Swifts' wings appear to beat the air alternately. The birds move in great curves, seldom alight, and drop suddenly into chimneys at night or when they wish to enter their nests.

Nest: A wall-pocket, built of sticks glued together and to the wall by a sticky saliva secreted by the swifts. During rainy weather the nest is sometimes loosened, and falls.

Eggs: White, like those of woodpeckers and some others laid in dark places.

Habitat: As swifts secure all of their food while on the wing and seldom alight, they have no habitat except the atmosphere and the hollow trees or chimneys in which they congregate at night, and where they nest. They do not perch on telegraph wires as swallows like to do.

Range: Breed in eastern North America, from southcentral Canada to the Gulf, and westward to the Plains; winter south of the United States.

Figure 34. Chimney swift.

Swifts have often been called "Chimney Swallows," but the name is a misnomer; they belong to an entirely different family. The breadth of wing and rapid flight, the weak feet and broad bills are, however, points of resemblance; the sooty appearance and lack of beautiful luster of plumage are points of difference. Then, too, swifts' tails are less like swallows' tails than they are like those of woodpeckers and creepers; the spiny tips are used as props against a perpendicular surface.

The following facts concerning swifts are taken from Eaton's "Birds of New York":

> "Nearly every village or city [in New York State] can boast at least one large chimney or church or schoolhouse that harbors multitudes of swifts every night late in summer. It is an interesting sight to watch these swifts as they wheel about such a chimney in the August and September evenings and, when the magic moment arrives, pour down its capacious mouth in a living cascade. It seems impossible for this species to perch, but it always alights on some perpendicular surface like the inside of a large hollow tree or the inner surface of a chimney or the perpendicular boards at the gable end of a barn or shed. In this position it sleeps, clinging with its sharp claws to the irregular surface and using its spiny tail as a support. The swift is seen abroad early in the morning and late in the afternoon, but in cloudy weather comes out at any time of day and evidently can see well in the bright sunlight, for it frequently hunts material for its nest during the brightest weather. They begin to construct the nest 182 in May or early June, the small twigs of which it is formed being broken from dead branches of some shade tree by the bird flying directly against the tip of the twig and snapping it off. The twigs are carried into the chimney and are cemented to the wall and to each other by a gelatinous substance secreted by the salivary glands of the bird itself. When completed, the nest is like a little semi-circular bracket slightly hollowed downward. The eggs are placed on this framework of twigs without lining.
>
> "In food the swift is wholly insectivorous, and does an immense amount of good destroying beetles, flies, and gnats, which he devours in countless multitudes. The chimney swift, as he darts by, frequently utters a rapid chipper something like the syllable *chip, chip, chip*, rapidly

> repeated, and I have heard a loud cheeping in the chimney, evidently uttered by the young birds. One of the earliest impressions of my boyhood was the curious roaring caused by the wings of parent swifts as they came and went from their nests at daybreak. This unfortunate habit of early rising has brought the chimney swift into bad repute in many civilized communities, ... closing chimneys against this beneficial bird."

In Major Charles Bendire's "Life Histories of American Birds" occur the following statements from Mr. Otto Widman regarding the nests and young of chimney swifts: "The setting parent shields the structure by habitually covering its base with the breast and pressing its head against the wall above. When disturbed, it hides below the nest, as do the young birds. They make a hissing noise, and always remain 2 or 3 feet below the mouth of the chimney [shaft], where they are fed by the parents until they are four weeks old.

> "Few birds are more devoted to their young than the Chimney Swift, and instances are recorded where the parent was seen to enter chimneys in burning houses, even after the entire roof was a mass of flames, preferring to perish with its offspring rather than to forsake them."

The Whip-Poor-Will

Goatsucker Family—Caprimulgidæ

Length: Nearly 10 inches; wings 7 inches long.

General Appearance: A *mottled brown bird with a narrow white band around throat, and white outer tail-feathers.*

> "He seems a lichen on a log,
> A dead leaf on the ground."

Male and Female: Soft brown, irregularly mottled and barred with black, buff, and white. Throat dark with a *narrow curve of white* in the male, and one of *buff* in the female. Beak short, slightly hooked, and very wide (1½ inches), with long bristles at the sides. Breast dark, belly white. Middle tail-feathers mottled brown; *half of six other tail-feathers white*, which are visible in flight. Female has narrower white tips to outer tail-feathers.

Figure 35. Whip-poor-will.

Note: *Whip'-poor-will, whip'-poor-will, whip'-poor-will*, uttered rapidly, monotonously, lugubriously, continuously. My sister counted 275 repetitions of his note given without a pause. To some people the sound is unendurable. When near the bird, I have heard him give a soft *chuck* between the repetition of the word *whip-poor-will*. He is associated in my mind with bright moonlight evenings, for it is then he is most vociferous. He sings, also, early in the morning.

Flight: Swift, yet noiseless; almost as uncanny as his note.

Habitat: In woods and open groves, where one may come upon him both at night and during the daytime sitting lengthwise on a log or branch instead of crosswise.

Nest: No nest is made, but two dull-colored, mottled eggs are laid on the ground or on dead leaves.

Range: Eastern North America. Breeds from southern Canada to the northern parts of Louisiana, Mississippi, and Georgia, and from the Plains eastward; winters from eastern South Carolina and the southern Gulf States to Central America. The chuck-will's-widow is a resident of our southeastern states; the poor-will of our western states.

The whip-poor-will is too interesting and useful a bird to be disregarded. He has been widely disliked and even superstitiously dreaded because of his weird notes. He is, however, of especial interest to scientists because of his nocturnal habits and his value as a destroyer of insects. Mr. Forbush calls him "an animated insect trap," with an "enormous mouth surrounded by long bristles which form a wide fringe about the yawning cavity."[97] The whip-poor-will is believed to be the greatest enemy of night-moths; he eats other insects, also, in great quantities.

The chuck-will's-widow is even more interesting than the whip-poor-will. Mr. W. L. McAtee writes of the bird as follows:

> "Like other species of its family, it lays only two eggs, which may be deposited almost anywhere on the forest floor, there being no nest. Intrusion on this spot usually results in the bird moving the eggs, which it carries in its mouth. Although the bird is only 12 inches long, the mouth fully extended forms an opening at least 2 by 3½ inches in size. It is but natural, therefore, that the bird should prey upon some of the largest

[97] "Useful Birds and Their Protection," E. H. Forbush, page 343.

insects. Not only 186 are large insects captured and swallowed, but even small birds, in two cases warblers.

"Despite the fact that the chuck-will's-widow occasionally devours small insectivorous birds, it must be reckoned a useful species. It is probable that birds are not deliberately sought, but that they are taken instinctively, as would be a moth or other large insect coming within reach of that capacious mouth."[98]

THE NIGHTHAWK

Goatsucker Family—Caprimulgidæ

Length: 10 inches; wings 7¾ inches.

General Appearance: A large dark bird, with a *white throat*, *a white band across the tail*, and *very long wings*, on each of which is a *large white spot or bull's-eye*, unfortunately a target, like the white rump of the flicker.

Male: Black above, mottled with buff and white; under parts lighter (becoming whitish), barred with black; throat with a tent-shaped white patch below the *very wide* bill; upper breast black; tail notched, a white band extending across it near the end except on the middle tail-feathers; wing with a conspicuous area of white about half-way between the curve and tip, when outspread.

Female: Throat buff; under parts buffy; no white on the tail.

Note: A loud *peeng-peeng*; uttered at frequent intervals while on the wing.

Flight: Very swift, with numerous and rapid changes of direction. The bird is very active at nightfall. It makes rapid descents not unlike those made by an airplane; it has a habit of dropping "like a bolt from the blue."

Habitat: The nighthawk is a "bird of the air" rather than of treetops or ground. It may be seen in cities flying above houses in search of its insect prey at sunset and during the night.

---

[98] Farmers' Bulletin 755, Biological Survey, U. S. Dept. of Agriculture.

Nest: No nest, but two speckled eggs are laid on the ground or on a roof where they are not easily discovered. Mr. Forbush says, "The nighthawk has deposited its 188 eggs on gravel roofs in cities for at least forty years and probably longer."

Figure 36. Nighthawk.

Young: Dr. F. H. Herrick tells us that the nestlings are "clothed in down" and "look like two little flattened balls of fluffy worsted of a dark cream-color mottled with brown."

Range: Eastern and central North America. Breeds from Manitoba, southern Quebec, New Brunswick, and Nova Scotia south to northern Louisiana, Mississippi, and Georgia, and from eastern North Dakota, Nebraska, and Kansas eastward; winters from the lowlands of South Carolina and southern parts of the Gulf States to British Honduras and Salvador.

The nighthawk is a remarkable bird. Because of its nocturnal habits, it has been regarded with superstitious awe. Erroneous ideas of it have been entertained, and it has received a name that belies it. It is not a *hawk* at all; it preys only on insects, not on chickens or small rodents.

Mr. W. L. McAtee writes:

> "Nighthawks are so expert in flight that no insects can escape them. They sweep up in their capacious mouths everything from the largest moths and dragon flies to the tiniest ants and gnats, and in this way sometimes gather most remarkable collections of insects. Several stomachs have contained fifty or more different kinds, and the numbers of individuals may run into the thousands. Nearly a fourth of the bird's total food consists of ants."[99] Professor Beal estimated that the stomachs of eighty-seven nighthawks which he examined "contained not less than twenty thousand ants, and these were not half of the insect contents."[100] 189 Mr. Forbush claims that the nighthawk "ranks next to the flicker in the destruction of ants, and it takes them when they are flying and about to propagate."[101]

It has a fondness for fireflies, also. Dr. Herrick made careful observation of the habits of nighthawks, and the manner of feeding their young. He writes of seeing a mother-bird "loaded with fireflies." He says:

---

[99] Farmers' Bulletin 755, Biological Survey, U. S. Dept. of Agriculture.
[100] Farmers' Bulletin 630, Biological Survey.
[101] "Useful Birds and Their Protection," E. H. Forbush, page 342.

"As her great mouth opened you beheld wide jaws and throat brilliantly illuminated like a spacious apartment all aglow with electricity. She made an electrical display at every utterance of her harsh *ke-ark*. Then standing over her young, with raised and quivering wings, she put her bill down into his throat and pumped him full. She then tucked the little one under her breast and began to brood. She repeated the performance, after which she settled down to brood as if for the night. This young bird was fed but twice each evening between the hours of eight and nine o'clock, and always, as I believe, by the female. It is quite probable that another feeding occurs also at dawn. The male would sometimes swoop down and once he sat by the chick for ten minutes after dusk. The task of feeding was borne by the mother."[102]

THE HOUSE WREN

Wren Family—Troglodytidæ

Length: About 4¾ inches.

Male and Female: Cinnamon-brown above, reddish-brown on the rump and tail. Back with fine indistinct bars; wings and tail with heavier bars; under parts grayish-white washed with brown, lighter on throat and breast; sides, and feathers under tail, barred with black; tail frequently held upright.

Notes: Sharp scolding notes.

Song: A sweet bubbling song. The notes are poured forth with joyous abandon and tireless energy.

Habitat: Near the homes of man preferably, though in the winter many house wrens are found in southern woods. They dart in and out of wood-piles and brush-heaps, run along walls and fences, and seek shrubbery, vines, and orchards.

---

[102] From "The Home Life of Wild Birds," by Francis H. Herrick; used with the permission of the author and his publisher, the G. P. Putnam's Sons.

Figure 37. House wren.

Nest: Of small sticks, lined with root-fibers or grasses, placed in a hollow of a tree, in a nesting-box, or some out-of-the-way place, such as a flower-pot, tin-can, discarded shoe, old hat, etc.

Range: Eastern North America. Breeds from southeastern Canada, eastern Wisconsin and Michigan, southward to Kentucky and Virginia; winters in eastern Texas, and in the South Atlantic and Gulf States.

Little "Jenny Wren" figured in our nursery tales and was one of the delights of our childhood, because of its diminutive size, its pert, cocked tail, its incessant activity, and its continuous chatter. No dull moments when a wren was nearby!

Its nesting-habits make it interesting to young and old. Though loyal to a nesting-locality, it will make its neat nest in a great variety of places, such as boxes, empty jars, small pails, or gourds, if placed conveniently, or in wren-houses.

Wrens are valiant defenders of their nests, but have been driven away from favorite nesting-places by quarrelsome English sparrows; consequently wrens are decreasing in number. Wren-houses with openings about an inch in diameter, too small for sparrows to enter, may help somewhat to check the decrease of these valuable insect-eating birds.

They are noisy little neighbors, a curious combination of joyousness and irritability. A pair of wrens that built a nest on the piazza of my brother's home spent so much time in scolding and quarreling that they were almost unendurable. One morning they disappeared; a few hours later my brother found the drowned body of the female in a rain-barrel. Whether it was accident, murder, or suicide, no one knew, but within twenty-four hours a pleasanter-tempered Lady Wren appeared, swept and garnished the home of her predecessor, and set up house-keeping. A larger measure of peace reigned thereafter.

As songsters, wrens are very remarkable for volume of sound, for sweetness of tone, and for extreme ecstasy. I remember wakening about sunrise one morning in early June, when the spring chorus was at its climax. For about an hour, I had the joy of listening to a bird-concert more wonderful than any I had ever heard. After a time I distinguished the voices of the various familiar birds. Loudest, clearest, and sweetest of all rang the voice of the smallest member of the choir—that of the tiny house wren.

THE RUBY-THROATED HUMMINGBIRD

Hummingbird Family—Trochilidæ

Length: About 3¾ inches; bill over ½ inch.

Male: Iridescent green above; gray below, with a glint of green, especially on the sides; wings and tail brown, with slight iridescence; throat brilliant ruby,—brownish in some lights; *tail forked.*

Female: Similar to male, but *without ruby on throat*, which is flecked with minute brownish spots; *tail-feathers of nearly even length*, outer feathers with white tips.

Note: No song—only a faint squeak.

Habitat: Open country; cultivated tracts of land, especially those overrun with vines; gardens, particularly those that contain trumpet-creepers and honey-suckles.

Nest: One of the most exquisite nests made. It is in the shape of a tiny cup, covered with lichens and lined with soft materials. It is frequently placed so high on a branch as to be difficult to distinguish from an excrescence on the bough. The eggs look like white beans.

Range: Eastern North America. Breeds from central Canada to the Gulf Coast and Florida; winters from central Florida and Louisiana through Southern Mexico and Central America to Panama.

Hummingbirds are rightly in a family by themselves—they are unique. They are the smallest of our birds, and yet they possess a power of flight unsurpassed. Mr. Forbush says:

> "The little body, divested of its feathers, is no larger than the end of one's finger, but the breast muscles which move the wings are enormous in proportion to the size of the bird. They form a large part of the entire trunk, and their power is such that they can vibrate the inch-long feathers of those little wings with such rapidity that the human eye can scarcely follow the bird when it is moved to rapid flight by fear or passion."[103]

---

[103] "Useful Birds and Their Protection," by E. H. Forbush, p. 241.

The wings do not seem to be made of feathers, but of gauze, like those of insects. I never really saw the feathers until I held a dead hummingbird in my hand. Its iridescent body seems made of burnished metal.

Figure 38. Ruby-throated humming-bird.

It is wonderful that so tiny a creature can wing its way from Central America to the heart of Canada. It seems to know no fear; it is quite able to defend itself with its long sharp bill. Mr. Forbush says: "The males fight with one another, and, secure in their unequalled powers of flight, they attack other and larger birds. When the Hummingbird says 'Go!' other birds stand not upon the order of their going, but go at once; while the little warrior sometimes accelerates their flight, for his sharp beak is a weapon not to be despised. Even the Kingbird goes when the war-like Hummer comes; the English Sparrow flees in terror; only the Woodpeckers stand their ground."[103]

Hummingbirds are not only fearless and pugnacious, but they are very inquisitive. Major Bendire says:

> "I once occupied quarters that were completely covered with trumpet-vines, and when these were in bloom the place fairly swarmed with Ruby-throats. They were exceedingly inquisitive, and often poised themselves before an open window and looked in my rooms, full of curiosity, their bright little eyes sparkling like black beads. I caught several—by simply putting my hand over them, 194 and while so imprisoned they never moved, and feigned death, but as soon as I opened my hand they were off like a flash. They seem to be especially partial to anything red."[104]

Their fondness for honey-producing flowers has caused many people to believe that they live upon nectar and ambrosia, like the gods of the Greeks, but the Biological Survey, has, by close observation, discovered that they do not visit flowers wholly for the purpose of gathering honey, but for obtaining also small insects that have been drowned in a welter of sweetness. Professor Beal has observed them "hovering in front of a cobweb, picking off insects and perhaps spiders entangled in the net. They have also been observed to capture their food on the wing, like flycatchers. Stomach examination shows that a considerable portion of their food consists of insects and spiders." Professor Beal continues:

> "Although hummingbirds are the smallest of the avian race, their stomachs are much smaller in proportion to their bodies than those of other birds, while their livers are much larger. This would indicate that these birds live to a considerable extent upon concentrated sweets, as stated above, and that the insects, spiders, etc., found in the stomachs do not represent by any means all their food."[105]

---

[104] "Life Histories of North American Birds," Maj. Chas. Bendire.
[105] Farmers' Bulletin 506, Biological Survey, U. S. Dept. of Agriculture.

A physician of my acquaintance owns a camp in the New Hampshire woods. A birch near his house was attacked by sapsuckers. Sap exuded plentifully and was eagerly sought by two red squirrels, a small swarm of bees, two sapsuckers, and seven hummingbirds. With his 195 glasses, the doctor observed the birds eating insects served in birch syrup.

Professor Beal reports having seen as many as one hundred hummingbirds "hovering about the flowers of a buckeye tree, and this number was maintained all day and for many days, though the individuals were going and coming all the time." Burroughs once saw a hummingbird take his morning bath in dewdrops.

There are about five hundred known species of hummingbird. They may be found in North and South America from Alaska to Patagonia. They are most numerous in northern South America, in Colombia and Ecuador. Seventeen species are found in our western and southwestern states, but only one, the Ruby-throat, lives in the East.

The Indigo-Bird or Indigo Bunting

Finch Family—Fringillidæ

Length: About 5½ inches.

Male: Head and throat deep, purplish blue, becoming lighter on back and above tail; wings and tail a brownish black, edged with blue. Winter plumage, brownish like the female, mottled with blue.

Female: Brown above, darker on wings and tail; *no streaks on back*; breast grayish, washed and *faintly streaked with brown*; belly lighter. The female resembles her sparrow relatives, but may be distinguished by a glint of blue in her tail and wings.

Call-note: A sharp *chip*.

Song: A burst of melody, somewhat like that of a canary, loud, clear, and sweet. It is not remarkable except that it may be heard during the middle of the day and during the heat of midsummer. The bird sings frequently from treetops.

Habitat: In “scrubby” pastures, along roadsides—in trees and bushes.

Range: Eastern North America. Breeds east of the Great Plains from North Dakota to New Brunswick, south from central Texas to Georgia; winters from southern Mexico to Panama.

The Indigo Bunting possesses a brilliant beauty and a sweet voice. A sight of him and his pretty brown mate brings a thrill of pleasure, but he holds no such place in our affections as does the true bluebird. He does not choose to nest close to human dwellings, but prefers overgrown pastures, not too much frequented, where he performs his good office of caterpillar-, canker worm-, and grasshopper-hunting, varying his diet with an abundance of weed seeds.

Figure 39. Indigo-bird.

The indigo-bird, the scarlet tanager, the goldfinch, and the Baltimore oriole are our most brilliant summer birds. Thoreau, in his "Notes on New England Birds" makes the following comment:

> "This is a splendid and marked bird, high-colored as is the tanager, looking strange in this latitude. Glowing indigo. It flits from the top of one bush to another, chirping as if anxious. Wilson says it sings, not like most other birds in the morning and evening chiefly, but also in the middle of the day. In this I notice it is like the tanager, the other fiery-plumaged bird. They seem to love the heat."

During August, the songs of the indigo-bird and red-eyed vireo may be heard along wooded roadsides, and are especially welcome because most birds are silent at that time.

The Baltimore Oriole

American-Blackbird Family—Icteridæ

Length: About 7½ inches.

Male: Head, throat, neck, and upper half of back black; breast, belly, shoulders, lower half of back and outer tail-feathers *brilliant orange*; wings black, many feathers edged with white; half of middle tail-feathers black; others largely orange; bill long, slender, sharp.

Female: Upper parts grayish-olive, washed with yellow and mottled with black on head and back; under parts, tail, and rump dull orange, paler at throat, which is sometimes marked with black; wings brown, barred with white.

Notes: A loud *whew-y*, or *whew*, uttered frequently and insistently, with a falling inflection. Orioles chatter noisily, also.

Song: A rich, melodious strain, very different in individuals, but alike in a liquid quality, and in frequency of utterance. For several successive years, two orioles returned to our elms and apple-trees in Cleveland. Their songs differed as decidedly from each other and from those of other orioles as the voices and enunciation of people vary.

Habitat: Elm and maple-shaded streets and orchards preferred in the springtime. After the nestlings are grown, orioles may be found in thickets or in the woods.

Nest: A hanging nest in the shape of a bag, usually suspended near the end of a bough. The female weaves the nest.

Range: Breeds from southern Canada and northern United States to the northern part of Texas, Louisiana, and Georgia, west to the Rocky Mts.; winters from southern Mexico to Colombia.

Figure 40. Baltimore oriole.

Orioles, with their brilliant plumage and beautiful song, belong to the somber-hued, unmusical blackbird family. They are truly "the flower of the flock,"—gorgeous tropical flowers, too. They invariably arouse interest and almost always great admiration. So dashing are they that they do not remain long enough near us to let us know them well or love them. They remind me of brilliant opera-singers, elegantly attired, who are followed by the eager eyes of a host of people.

So many poets and writers of prose have sung the praise of orioles that it surprised me to learn that neither Thoreau nor Burroughs admired them. Thoreau wrote:

> "Two gold robins; they chatter like blackbirds; the fire bursts forth on their backs when they lift their wings.... But the note is not melodious and rich. It is at most a clear tone."[106]

Burroughs said:

> "I have no use for the oriole. He has not one musical note, and in grape time his bill is red, or purple, with the blood of our grapes."[107]

A grape-eating propensity is not a trait common to orioles, according to Professor Beal's report of their food habits. He says:

> "Brilliancy of plumage, sweetness of song, and food habits to which no exception can be taken are characteristics of the Baltimore oriole. During the stay of the oriole in the United States, vegetable matter amounts to only a little more than 16 per cent. of its food, so that the possibility of its doing much damage to crops is very limited. The bird is accused of eating peas to a 200 considerable extent, but remains of such were found in only two cases. One writer says that it damages grapes, but none were found in the stomachs."[108]

---

[106] From "Notes on New England Birds," by H. D. Thoreau.
[107] From "Under The Maples," by John Burroughs.
[108] Farmers' Bulletin 630, Biological Survey, U. S. Dept. of Agriculture.

Professor Beal lists caterpillars, beetles, bugs, ants, wasps, grasshoppers, and some spiders as the "fare of the oriole."

The nest and nesting habits of these birds are unusually interesting. In Eaton's "Birds of New York" occurs the following description:

> "The female is an ideal mother, defending her young with great courage and caring for them in all kinds of weather. The young, however, are not such ideal offspring as she ought to expect. From the time they begin to feather out until several days after they have left the nest, they keep up a continual cry for food. In this way they are unquestionably located by many predaceous animals and thereby destroyed. The young orioles are usually out of the nest from the 20th of June to the 5th of July [in New York State], and are very soon led away by the old birds into the woods, groves, and dense hedgerows. Then we hear no more of the oriole's song until the latter days of August or the first week in September, when, after the autumn molt has been completed, the males frequently burst into melody for a few days before departing for their winter home.
>
> "As everyone knows, the oriole builds a pensile nest, usually suspending it from the drooping branches of an elm tree, soft maple, apple tree, or in fact, any tree, though his preference seems to be for the elm. The main construction materials used by the oriole are gray plant-fibers, especially those from the outside of milkweed stalks, waste packing-cord and horsehair; sometimes pieces of 201 rags and paper are discovered in the nest, but it is almost without exception a grayish bag as it appears from the outside, and is lined principally with horsehairs and softer materials, making a thick felted gourd-shaped structure."

One morning this past May when the heat was unseasonable and overpowering, an oriole was observed fluttering anxiously near the nest where his mate sat on her eggs. The foliage had not developed sufficiently to shade her, so he alighted on the nest, a claw on either side of the cup-like opening. There he stood astride for the greater part of the day and protected her devotedly, like a chivalrous knight of old.

The Orchard Oriole

American Blackbird Family—Icteridæ

Length: About 7 inches.

Adult Male: Breeding Plumage: Head, throat, neck, and upper half of back black; breast, belly, shoulders, lower half of back a bright chestnut brown; wings and tail dark brown; wing-feathers tipped or edged with white, forming a bar across wing. The winter plumage is different from the breeding plumage; the male passes through several changes as he matures.

Female: Olive-green above, darkest on head and back, dull yellow below; wing-feathers tipped with white, forming two bars across wing; tail olive-green.

Immature Male: Like female, the first autumn; the next spring, he has a black throat; the chestnut plumage develops later.

Notes and Song: Similar to those of the Baltimore oriole. Song clear and melodious; tones possibly not quite so *mellow* as those of its relatives, but sweeter.

Habitat: Orchards and shade trees.

Nest: A pensile nest, but shorter and more firmly attached than that of the Baltimore oriole.

Range: Eastern North America. Breeds from northern United States, southern Canada, and central New York, south to northern Florida and the Gulf Coast, west to Texas, central Nebraska, and western Kansas; winters from southern Mexico to northern Colombia. Not common in Massachusetts.

The markings of the Orchard Oriole are similar to those of the more brilliant and striking Baltimore Oriole, but its coloring more nearly resembles that of the towhee. Like its cousin, it is arboreal, while the towhee is a ground bird.

The orchard oriole is more shy than the Baltimore oriole and is less well known. It is, however, very active and restless,—indefatigable in its quest for insects.

It has a better reputation than most members of the blackbird family. Major Bendire says that it would be difficult to find a bird that does more good and less harm than the orchard oriole, and that it should be fully protected.

Figure 41. Orchard oriole.

THE SCARLET TANAGER

Tanager Family—Tangaridæ

Length: About 7 inches

General Appearance: A bright scarlet body, with black wings and tail; no crest.

Male: Scarlet and black in breeding plumage; after the molt, olive and yellow, with black wings and tail; wings white underneath. The male does not acquire red plumage until the second year. While molting, the adult male has irregular patches of olive and yellow mixed with his red feathers, giving a curious effect.

Female: Olive-green above; yellowish-olive below, brightest on throat; wings and tail dark gray, washed with olive. She is very effectively protected by her coloring.

Note: Call-note *chip-chur*, very distinct and reasonably loud.

Song: A warble, full, rich, and pleasing, but not varied; sufficiently like the songs of the robin and the rose-breasted grosbeak to make identification difficult for a beginner. The frequent *chip-chur* betrays the tanager's presence.

Habitat: Dense groves of hard-wood trees, especially those containing oaks. Mr. Forbush calls the tanager "the appointed guardian of the oaks." The bird is found in parks and on well-wooded estates, as well as in the deep woods.

Range: Eastern North America and northern South America. Breeds in southern Canada as far west as the Plains, and in the United States to southern Kansas, northern Arkansas, Tennessee, northern Georgia, and the mountains of Virginia and South Carolina; winters from Colombia to Bolivia and Peru.

These "black-winged redbirds" are occasionally mistaken by novices for cardinals, but the dusky wings and tail, and the absence of a crest differentiate them. Then, too, the scarlet of their coats is of a different shade of red.

Figure 42. Scarlet tanager.

Their cousins, the summer tanagers, denizens of southeastern United States and occasional residents of the North, resemble cardinals more closely. Both have a nearly uniform rose-red plumage, but the summer tanager has brownish wings edged with red, and *no crest*.

The beauty of male tanagers has caused them to be eagerly sought in the past. I have childish memories of their scarlet bodies decorating the hats of thoughtless women, and I blush to confess a feeling of envy rather than regret at the wicked slaughter. Audubon Societies have done much to change public sentiment and put a stop to barbarous practices.

Never shall I forget the breathless joy I felt when, grown to young womanhood, I first saw a tanager's vivid beauty gleaming against the almost black-green foliage of a dense grove. I think that I remember every tanager which I have since seen, as well as each lovely setting that enhanced his gorgeous coloring. A glimpse of one marks a red-letter day. Twice I have seen two males at once, in company with a rose-breasted grosbeak—all singing; memorable experiences.

The western tanager, with his yellow body and crown, his red "face," black back and tail, and yellow and black wings, appeared before me one day in the noble woods that crown Glacier Point in the Yosemite Valley. I felt that his beauty, like that of his eastern relatives, was his "excuse for being." He does not enjoy quite so good 206 a reputation as do other tanagers, because he has a taste for fruit—almost as reprehensible as horse- or cattle-stealing in the west.

Tanagers, however, are valuable insect-destroyers. Our brilliant species deserves our whole-hearted protection, not only for aesthetic, but also for economic reasons.

The Rose-Breasted Grosbeak

Finch Family—Fringillidæ

Length: A little over 8 inches.

General Appearance: A black and white bird, with a *rose-colored breast* and *heavy, flesh-colored beak.*

Male: Head, throat, and back black; rump and under parts white, except *on breast* and *under wings*, which are a beautiful *rose-red*; wings black, with bars and patches of white; tail black; outer feathers with white tips to their inner webs. The winter plumage is slightly different from the summer plumage.

Figure 43. Rose-breasted grosbeak.

Female: A soft grayish-brown, streaked with white, buff, and gray; under parts light buff, faintly streaked with brown; head brown; a buff streak through center of the crown, a white streak over the eye; wings and tail grayish-brown, some of the wing-feathers tipped with white; yellow under wings instead of rose.

Note: A sharp *tsick, tsick*.

Song: A rich, beautiful warble, somewhat like that of the robin and tanager, but more joyous than either. It possesses a purer, more liquid quality. The song is remarkable, also, in that it may be heard at night, and at midday.

Habitat: Woodlands and thickets, fields and gardens. This grosbeak frequents also the shade trees of large estates and suburban streets.

Nest: Large and loosely constructed, made of twigs, grasses, and root-fibers, and placed from five to twenty feet from the ground.

Eggs: Pale blue, spotted with brown or purple. The male takes his turn at sitting on the eggs.

Range: Eastern North America and northern South America. Breeds from southern Canada south to Kansas, Missouri, Ohio, New Jersey, and in the mountains of northern Georgia; winters from southern Mexico to Colombia and Ecuador.

So beautiful is the rose-breasted grosbeak and so melodious his song that he invariably attracts attention. Upon close acquaintance, he reveals many interesting habits and delightful traits. He is so useful that he reminds one of the occasional rare person who combines practical qualities with beauty of form and face and unusual gifts.

He is one of our most beneficial birds. Occasionally he partakes of cultivated fruit and devours green peas, but the slight mischief he is guilty of is greatly overbalanced by the good he does. So fond is he of the Colorado potato beetle that in some localities he is called the "potato-bug bird."[109] Professor Beal tells of watching grosbeaks near a potato-patch that was nearly riddled by these destructive insects. He saw the parent-birds visit the field repeatedly, and then bring their young when able to fly. The brood perched in a row on the top rail of the fence, and were fed so frequently that in a few days the potato-bugs had entirely disappeared. The crop was saved.

Grosbeaks appear to lead unusually happy domestic lives. Though the males fight for their mates, they guard them and their young with great

---

[109] Farmers' Bulletin 513, Biological Survey, U. S. Dept. of Agriculture.

devotion. They not only utter low sweet notes to the mother-bird as she broods, but quite frequently take her place on the nest.

My sister tells of hearing a rose-breast's song in a maple 209 grove, and of searching diligently for the singer. She located the tree from which the sound proceeded, and waited patiently to see him "gaily flit from bough to bough"; but no bird came into view. She went around the tree until, to her delight, she discovered him sitting on the nest, only a few feet from where she stood. He stopped singing when he saw her, but showed neither surprise nor fear, and resumed his song after she went away. She realized that she had had an unusually rare privilege.

To hear a grosbeak's song at night is an experience similar to that of listening to a nightingale in Europe, or to a mockingbird in our South or West, singing by moonlight.

The Blue Grosbeak

Finch Family—Fringillidæ

Length: 7 inches; indigo bunting, 5½ inches.

Male: Body a deep blue, almost black on the back; chin and cheeks black; bill heavy; tail black, edged with blue; wings black, tipped with bright brown, giving the effect of one broad and one narrow wing-bar. Winter plumage, rusty brown mottled with blue.

Female: Grayish-brown above, more or less washed with blue; wings brown, barred with buff; under parts washed with buff.

Song: A sweet grosbeak warble.

Range: Eastern North America. Breeds from Missouri, southern Illinois and Maryland, south to eastern Texas, and northern Florida; accidental in Wisconsin, New England, the Maritime Provinces, and Cuba; winters in Yucatan and Honduras.

The Blue Grosbeak resembles its smaller relative, the indigo bunting, but it has a larger, darker body, a heavier bill, and brown-tipped wing feathers. It is more nearly the size of a cowbird than of the indigo-bird. It may be found in thickets similar to those frequented by its small blue relative.

It is a bird of the southeastern part of the United States, but occasionally strays northward.

THE EVENING GROSBEAK

Finch Family—Fringillidæ

Length: 8 inches; 3 inches larger than the goldfinch.

Male: Forehead bright yellow; crown of head black; body olive-brown, with yellow on shoulders, rump, and belly; wings black and white; tail forked, black; bill heavy and yellowish.

Female: Brownish-gray, tinged with yellow underneath; wings black and white; forked tail black, tipped with white.

Range: Central North America. Breeds in western Alberta; winters in the interior of North America east of the Rocky Mts., more or less irregularly in southern Missouri, Kentucky, Ohio, eastern Pennsylvania, New York, New Jersey, New England, and Quebec.

A sight of this handsome bird is an event in the East, and arouses great interest in people who know how rare it is. Five were seen near Washington in early April of this year, and were hailed with enthusiasm. It is a common resident of our Northwest, though it wanders in flocks to the East occasionally.

It looks like a large goldfinch, though it is a less brilliant yellow, has larger patches of white on its wings and wears its dark cap back on its head, above its yellow forehead, instead of pulled down to its eyes and bill. It blends perfectly with the yellows and olive-browns of some of our western landscapes.

It feeds on berries, seeds, and insects. It becomes very tame.

The Black-Headed Grosbeak

Finch Family—Fringillidæ

The Black-headed Grosbeak has *cinnamon-brown* upper parts, breast, band about the neck, and rump; yellow belly, black head, wings, and tail; wings with two white bars and a white patch; tail with white tips. Female brownish-black and buff above; under parts tawny and yellow, streaked with dark; chin, sides of throat, and line over eye whitish.

"The Black-headed Grosbeak takes the place in the West of the rosebreast of the East, and, like it, is a fine songster. Like it, also, the blackhead readily resorts to orchards and gardens and is common in agricultural districts. The bird has a very powerful bill and easily crushes or cuts into the firmest fruit. It feeds upon cherries, apricots, and other fruits, and also does some damage to peas and beans, but it is so active a foe of certain horticultural pests that we can afford to overlook its faults.... It eats scale insects, cankerworms, codling moths, and many flower beetles, which do incalculable damage to cultivated flowers and to ripe fruit."[110]

The Bobolink

American Blackbird Family—Icteridæ

Length: A little over 7 inches.

Male: *Spring* or *Breeding plumage*: Crown, sides of head, throat, and other under parts black; *back of head and neck light yellow*; upper half of back black, streaked with creamy white; *lower half of back, rump, and shoulders white*; wings black, some of the feathers tipped with buff; tail black, the feathers pointed. Many birds have dark upper parts and light

---

[110] Farmers' Bulletin 513, Biological Survey, Dr. Henry W. Henshaw.

breasts; the bobolink wears his bright breast upon his back during the summer. In the fall, he resembles the female.

Female: Olive-brown and light yellow above, with black streaks; head with olive-brown and light yellow stripes; under parts pale yellow; wings and tail brown.

Notes: A tinkling *ding-ding*, not unlike the sound of a bell; likewise a chirp.

Song: A bubbling song, full of ecstasy and abandon. It is one of the most delightful songs of the later migrants.

Habitat: While in the North, the bobolink inhabits our fields and meadows, where he "swings on brier and weed." In the fall, he frequents the rice-fields of our southern states on his way to South America, and does so much harm that he is dreaded and hated.

Range: North and South America. Breeds mainly from the plains of south-central Canada to Nevada, Utah, northern Missouri, Illinois, Indiana, Ohio, West Virginia, Pennsylvania, and New Jersey; winters in South America, to southern Brazil, Paraguay, and Bolivia.

Had Robert Louis Stevenson written the biography of a bobolink, he might have given him the names of his immortal Dr. Jekyll and Mr. Hyde, for the bird seems to possess a dual nature, and to bear totally different reputations in the North and the South. When he visits Canada and northern United States in May, dressed in his gay wedding finery, he is greeted with joy. Few more delightful birds are to be found than this attractive, happy-hearted singer against whom no reproaches are registered in the North.

His song has been a favorite theme for poets and nature-writers. Thoreau wrote:

> "One or two notes globe themselves and fall in bubbles from his teeming throat. It is as if he touched his harp within a vase of liquid melody, and when he lifted it out, the notes fell like bubbles from the strings. Methinks they are the most liquidly sweet and melodious sounds I ever heard."[111]

---

[111] From "Notes on New England Birds," by Thoreau, page 246.

Figure 44. Bobolink.

The bobolink's habits in the North are almost beyond reproach. Professor Beal writes:

"In New England there are few birds about which so much romance clusters as this rollicking songster, naturally associated with the June meadows; but in the South there are none on whose head so many maledictions have been heaped on account of its fondness for rice. During its sojourn in the Northern States it feeds mainly upon insects and seeds of useless plants; but while rearing its young, insects constitute its chief food, and almost the exclusive diet of its brood. After the young are able to fly, the whole family gathers 214 into a small flock and begins to live almost entirely upon vegetable food. This consists for the most part of weed seeds, since in the North these birds do not appear to attack grain to any extent. They eat a few oats."[112]

Dr. Henshaw adds:

"When the young are well on the wing, they gather in flocks with the parent birds and gradually move southward, being then generally known as *reed-birds*. They reach the ricefields of the Carolinas about August 20, when the rice is in the milk. Then until the birds depart for South America, planters and birds fight for the crop, and in spite of constant watchfulness and innumerable devices for scaring the birds a loss of 10 per cent. of the rice is the usual result."[113]

Major Bendire, in his "Life Histories of North American Birds," quotes a letter from Capt. W. M. Hazzard, a large rice-grower of South Carolina, written concerning the warfare waged against these *ricebirds*:

"The Bobolinks make their appearance here during the latter part of April. At that season, their plumage is white and black, and they sing merrily when at rest. Their flight is always at night. In the evening there are none. In the morning their appearance is heralded by the popping of whips and firing of musketry by the bird-minders in their efforts to keep the birds from pulling up the young rice. This warfare is kept up incessantly until about the 25th of May, when they suddenly disappear at night. Their next appearance is in a dark yellow plumage, as the *Ricebird*. There is no song at this time, but instead a chirp which means ruin to any rice found in the milk. My plantation record will show that for the 215 past ten years, except when prevented by stormy south or southwest winds, the Ricebirds have come punctually on the night of the 21st of August, apparently coming from seaward. All night their chirp can be

---

[112] Farmers' Bulletin 630, U. S. Dept. of Agriculture, Biological Survey.
[113] Farmers' Bulletin 513, U. S. Dept. of Agriculture, Biological Survey.

heard passing over our summer homes on South Island, which is situated 6 miles to the east of our rice plantations, in full view of the ocean. Curious to say, we have never seen this flight during the day. During the nights of August 21, 22, 23, and 24, millions of these birds make their appearance and settle in the ricefields. From the 21st of August to the 25th of September our every effort is made to save the crop. Men, boys, and women with guns and ammunition, are posted.... The firing commences at dawn and is kept up till sunset.... If from any cause there is a check to the crop during its growth which prevents the grain from being hard, but in milky condition, the destruction of such fields is complete, it not paying to cut and bring the rice out of the field.... I consider these birds as destructive to rice as the caterpillar is to cotton, with this difference, that these Ricebirds never fail to come."

THE GOLDFINCH

Finch Family—Fringillidæ

Length: About 5 inches.

Male: *Spring* and *summer plumage—body* and *shoulders bright yellow*; *crown black*; *wings* and *tail*, *black* and *white*; tail forked; feathers above tail, gray. *Winter plumage*—olive-brown back; throat, breast, and shoulders yellow; wings black and white.

Female: Olive-brown above; dull yellow below; wings and tail a dull black; white bars on wings, tail white-tipped; shoulders olive-green; grayish above tail. *No black on crown.*

Notes: An unusually sweet chirp or call-note like that of a canary, *who-ee'*, with a rising inflection; a flight-note, *per-chick'ory*, given as the goldfinch bounds through the air; a number of gentle little twittering sounds, for these birds are very social and communicative.

Song: A rapid outpouring of notes in a wild, sweet, canary-like strain.

Flight: In great waves or undulations.

Habitat: Fields and gardens, or wherever its favorite food may be obtained.

Nest: In bushes or trees; made of soft grasses or fibers, and lined with thistledown.

Range: Eastern North America. Breeds from south-central Canada to Oklahoma, Arkansas, and northern Georgia; winters over most of its breeding range and south to the Gulf Coast.

Figure 45. Goldfinch.

In winter, the goldfinch may be distinguished from others of the finch or sparrow family by its undulating flight, its flight-note, *per-chick'ory*, and its call-note. Its black and white wings and tail are also distinctive. It is found in flocks during the winter season.

The Goldfinch or "Wild Canary" is one of our best-loved birds. The beauty of the male's coloring, the sweetness of his voice, the joyousness of his nature have won him many friends.

John Burroughs wrote:

> "The goldfinch has many pretty ways. So far as my knowledge goes, he is not capable of one harsh note. His tones are either joyous or plaintive. In his spring reunions they are joyous. In the peculiar flight song in which he indulges in the mating season, beating the air vertically with his round open wings, his tones are fairly ecstatic. His call to his mate when she is brooding, and when he circles about her in that long, billowy flight, the crests of his airy waves being thirty or forty feet apart, calling, 'Perchic-o-pee, perchic-o-pee,' as if he were saying, 'For love of thee, for love of thee,' and she calling back, 'Yes, dearie; yes, dearie'—his tones at such times express contentment and reassurance.
>
> "When any of his natural enemies appear—a hawk, a cat, a jay,—his tones are plaintive in sorrow and not in anger.
>
> "When with his mate he leads their brood about the August thistles, the young call in a similar tone. When in July the nesting has begun, the female talks the prettiest 'baby talk' to her mate as he feeds her. The nest-building rarely begins till thistledown can be had, so literally are all the ways of this darling bird ways of softness and gentleness. The nest is a thick, soft, warm structure, securely fastened in the fork of a maple or an apple-tree."[114]

The fondness of goldfinches for the seeds of thistles has given them the name of thistle-birds. While they eat insects during the summer, they are especially useful as seed-destroyers. At Marshall Hall, Md., Dr. Judd observed them eating their first fresh supply in the spring from dandelions; in June, they ate the seeds of the field daisy; in July, of the purple aster and

[114] From "Under The Maples," by John Burroughs; page 42.

wild carrot. Thistles and wild lettuce were feasted upon during August; while in September the troublesome beggar-tick and ragweed were eagerly sought. At one time Dr. Judd counted a flock of three hundred goldfinches busily stripping seeds from a rank growth of the latter weed; he discovered them, also, devouring seeds of the trumpet-creeper. They are invaluable aids to a farmer; the only fault of which they can be accused is that of "pilfering" sunflower seeds. The presence of sunflowers in a garden is likely to attract goldfinches, just as trumpet-creeper blossoms lure hummingbirds.

I recall a lovely garden in which I spent many pleasant hours one summer, happy in its beauty and fragrance, and in the companionship of bird visitors. Near my accustomed seat grew a clump of sunflowers, often sought by goldfinches. The black and gold of their plumage made a pretty sight against the yellow petals and dark centers of the great flowers. I remember one little bird that fluttered among the golden petals, too busy singing to eat for a time.

Two bird-hunting cats haunted the garden. I took a malicious pleasure in driving them away, because their ignorant, parsimonious owner had informed me that she kept them locked up while her chickens were young, so the cats wouldn't catch them. She didn't care how many 219 birds were killed, for then she wouldn't be obliged to feed the prowlers. The goldfinches soon learned that when I was there they could feast in safety. More than once when I was in the house or on the porch I would hear their alarm cry of Dé-de? dé-de? sound from a maple near the piazza, plainly calling for my aid. When I went out to the garden and drove away their feline foes, the cries would cease. The angry owner of the cats, who dared not remonstrate further with me, cut down the sunflowers!

My most beautiful memory of goldfinches is associated with one of their spring mating-festivals. My sister and I had read Burroughs's description of these love-feasts, so we were prepared to understand what the unusual chorus meant. The sweet call-notes of the males, interspersed with rapturous bursts of melody and frequent flutterings met with quick response from the olive-and-gold females, who chirped and said "Yes" with a joy pleasant to see! It is impossible to convey adequately any idea of the exquisite tenderness of their voices, of the absence of quarreling and jealousies,—of the perfect harmony of the proceeding. I can only wish that every person who loves birds might some time have the pleasure of a similar experience.

The Catbird

Mockingbird Family—Mimidæ

Length: Nearly 9 inches.

Male and Female: A slender, long-tailed, gray bird, with a *black crown* and *tail*, and *chestnut-brown feathers under the tail*; breast somewhat paler than back; bill slightly curved.

Note: A soft *wă*, not unlike the mew of a kitten.

Song: A delightful warble—soft, sweet, and musical, though it is occasionally interspersed with the catlike noise *wă*, and with sounds of mimicry. Catbirds are sometimes called northern mockingbirds.

Habitat: Tangled thickets preferred. Fruit trees, berry-patches, and garden-shrubbery are also sought.

Nest: A veritable scrap-basket made of twigs, leaves, grasses, plant-fibers and rootlets, with paper sometimes interwoven. One nest that I examined contained a scrap from a torn letter and a fragment of a sermon from a newspaper. Several tell-tale cherry-stones lay on the bottom, circumstantial evidence of theft.

Eggs: A lovely greenish-blue, not unlike those of the robin.

Range: A common bird of eastern North America, from central Canada to the Gulf and northern Florida. It is found in the northwestern part of the U. S. and winters in our southern states and in Central America.

The catbird is well-named. It is the color of a Maltese cat, is sleek and agile, and in movement quiet and stealthy. Its mew is so like that of a kitten as to be confusing to the uninitiated. I recall the frantic barking of our small dog at a catbird that she heard in the shrubbery one day. It was difficult to convince her that one of her hated foes, a cat, was not the author of the sound that always infuriated her.

Figure 46. Catbird.

Though catbirds possess little claim to beauty, they seem to be vain and appear always to be doing something to attract attention. They are in constant motion—twitching their tails, jerking their bodies, and making their gentle, inane "cat-calls."

I once had an amusing experience with a catbird. I had seated myself near a thicket in which a Maryland Yellow-throat was flitting. Hoping to beguile him from the shrubbery and thus afford myself a better view of him, I gave his song repeatedly—"Witch-a-tee-o, witch-a-tee-o." A catbird on the fence-rail behind the thicket was flirting his tail, looking knowingly at me, and giving his call repeatedly. I paid no attention to him, and continued to say "Witch-a-tee-o." It was not long before he, too, warbled "Witch-a-tee-o." Whether he did it from his love of mimicry or from a desire to be noticed, I shall never know, but his bearing was, "*Now* will you pay some attention to *me*!"

Catbirds are in disfavor among the growers of cherries and berries, both wild and cultivated; they make havoc in strawberry-beds. Mr. Forbush reports that their depredations vary in different localities. He claims that in spite of their fruit-stealing propensities they deserve protection in Massachusetts, because they devour locusts, cankerworms, and the caterpillars of various moths, most important being those of the gypsy and brown-tail moths.

In the Biological Survey Bulletin "Fifty Common Birds of Farm and Orchard" (No. 513) the following statements 222 about the catbird are made:

> "Half of its food consists of fruit, and the cultivated crops most often injured are cherries, strawberries, raspberries, and blackberries. Beetles, ants, crickets, and grasshoppers are the most important element of its animal food. The bird is known to attack a few pests such as cutworms, leaf beetles, clover-root curculio, and the periodical cicada, but the good it does in this way probably does not pay for the fruit it steals. The extent to which it should be protected may perhaps be left to the individual cultivator; that is, it should be made lawful to destroy catbirds that are doing manifest damage to crops."

Dr. Judd found that catbirds fed their young almost entirely on insects; he therefore scored a point in their favor. Their bravery in defense of their nest and their young is well known.

Burroughs tells an unusual anecdote about a catbird as follows:

> "A friend of mine who had a summer home on one of the trout-streams of the Catskills discovered that the catbird was fond of butter,

and she soon had one of the birds coming every day to the dining-room, perching on the back of the chair, and receiving its morsel of butter from a fork held in the mistress's hand. I think the butter was unsalted. My friend was convinced after three years that the same pair of birds returned to her each year because each season the male came promptly for his butter."[115]

Many other incidents might be related concerning this interesting bird,—of its unusual intelligence and its remarkable power of mimicry. One catbird in Tennessee learned to imitate the songs of all the birds that nested 223 near him. His rendering of the red-eyed vireo's song was as good as that of the vireo himself. His listeners felt that it was wearisome enough to have the red-eye preaching constantly, but to have the catbird reiterating it was more than they could endure.

THE BROWN THRASHER

Mockingbird Family—Mimidæ

Length: About 11 inches, larger than the robin; tail 5 inches long.

General Appearance: A large bird with a *bright brown back, white breast streaked with brownish-black*, and a *very long tail* which is moved or "thrashed" about incessantly.

Male and Female: Reddish-brown above; white underneath, becoming buff after the August molt; throat indistinctly marked with dark streaks; breast and sides *heavily streaked*; wings with two indistinct white bars; tail almost half the length of the bird; bill long (about 1 inch), sharp and curving.

Notes: A "smacking" sound and a sharp *whew*.

Song: A loud, clear, beautiful song. It consists of several phrases, each composed of two or more similar notes. Thoreau interpreted it as follows:

---

[115] From "Under the Maples," by John Burroughs, page 66.

"*cherruit, cherruit, cherruit; go ahead, go ahead; give it to him, give it to him.*"[116] The song is generally sung from the tops of trees or bushes.

Habitat: Like the catbird, the thrasher is found frequently in shrubbery, where it scratches among dead leaves for its food. Its brown color protects it admirably.

Nest: Made of twigs, leaves, and root-fibers, placed in thickets or on the ground.

Eggs: White, evenly speckled with fine brown spots.

Food: Wild fruit and berries (30 kinds), and insects, especially beetles and caterpillars. Professor Beal says: "The farmer has nothing to fear from depredations on fruit or grain by the brown thrasher. The bird is a resident of groves and swamps rather than of orchards and gardens."[117]

Range: Eastern United States and southern Canada, westward to the Rocky Mts.; winters in south-eastern United States.

Because of his brown color and his speckled breast, the Brown Thrasher has often been erroneously called the *Brown Thrush*. Careful observation reveals many points of difference. He is three or four inches longer than our common thrushes—in fact, his tail alone is only about 2½ inches shorter than the entire body of the veery or the hermit thrush; his bill is almost four times as long as theirs and is decidedly curved. Instead of dark, thrush-like eyes, he has pale yellow ones that give him an uncanny appearance.

He is not a dweller in woods, but, like the catbird, prefers thickets. Burroughs says: "The furtive and stealthy manners of the catbird contrast strongly with the frank open manners of the thrushes. Its cousin the brown thrasher goes skulking about in much the same way, flirting from bush to bush like a culprit escaping from justice. But he does love to sing from the April treetops where all the world may see and hear, if said world does not come too near."[118]

---

[116] From "Notes on New England Birds," by H. D. Thoreau, p. 361.
[117] Farmers' Bulletin 755, Biological Survey, U. S. Dept. of Agriculture.
[118] From "Under the Maples," by John Burroughs, p. 67.

Figure 47. Brown thrasher.

His song is a brilliant, delightful performance, admirable in technique, but lacking in a quality of tone that moves the heart. It is often of long duration. One May afternoon, I heard a thrasher singing so long that I was moved to time him. He sang without stopping for fifteen 226 minutes by my watch, and his entire song must have lasted nearly half an hour.

The brown thrasher, like the other members of his family, has power of mimicry. In the north, he is sometimes called the "Northern Mocker"; in some regions where he and the mockingbird both live, he is known as the "Sandy Mocker." There is sufficient similarity in the songs of the catbird, the thrasher, and the mockingbird to make a listener pause a moment to distinguish them when in a locality where the three birds are to be found. The catbird's mew betrays him; the thrasher's song is more brilliant and sustained; the mocker's more varied. Thoreau says, "The thrasher has a sort of laugh in his strain that the catbird has not."[119] His song resembles decidedly that of the English thrush, famed in poetry. Browning's description of the latter is equally applicable to our thrasher:

> "He sings each song twice over,
> Lest you should think he never could recapture
> That first fine careless rapture."

The Mockingbird

Mockingbird Family—Mimidæ

Length: About 10 inches; an inch longer than the catbird and an inch shorter than the thrasher; tail about 5 inches long.

Male and Female: A long, slender, brownish-gray bird, with grayish-white under parts; wings and tail dark brown; wings with two white bars and *white patches that are conspicuous in flight*; middle tail-feathers brown, *outer feathers white, others partly white*. The female frequently has less white than the male.

Notes: A great variety. Some mockingbirds seem to possess unlimited powers of mimicry; others have far less ability to reproduce sounds.

Song: A sweet, delightful melody, sung in pure liquid tones and with ease and assurance, as though the birds were conscious of their power.

---

[119] From "Notes on New England Birds," Thoreau, p. 361.

They are probably the most famous songsters of America. Sidney Lanier, Walt Whitman, and other poets have written well-known poems in their praise, while Roosevelt and many other prose-writers have added their encomiums.

Figure 48. Mockingbird.

Habitat: Near the haunts of man, in gardens, parks, tree-shaded streets, and groves.

Range: Southeastern United States chiefly from eastern Nebraska, southern Iowa, Illinois, Indiana, Ohio, and Maryland, south to eastern Texas, southern Florida and the Bahamas; occasional in New York and Massachusetts, though a number of records have been made near Boston; accidental in Wisconsin, Ontario, Maine, and Nova Scotia; introduced into Bermuda.

The western mockingbird is found in California, southern Wyoming, northwestern Nebraska, and western Kansas, south to Mexico and Lower California. It has a longer tail and wings than the eastern species, and is a paler gray.

Nuttall called the Mockingbird "the unrivalled Orpheus of the forest, and the natural wonder of America." His voice certainly has power to "soothe the savage breast," to interest the mind because of the varied range and remarkable technique, and to uplift the soul, especially when heard in the stillness and beauty of a moonlight night.

There is great difference of opinion regarding the "mocker." He is more loved and admired in the South than in the West, and is regarded with pride as worthy to be called the nightingale of America. Most writers have sung his praises, but occasionally someone regards him with disfavor because of his habit of interlarding his beautiful song with curious and disagreeable sounds. Wilson Flagg says, "He often brings his tiresome extravaganzas to a magnificent climax of melody and as frequently concludes an inimitable chant with a most contemptible bathos."[120]

The power of mimicry varies with different individuals. In a brief interval of time, one bird may imitate a woodpecker, a phœbe, a wren, a jay, or a cardinal, so as to deceive most listeners. He may produce the sound made by the popping of a cork or the buzzing of a saw; the next moment he may scream like a hawk to frighten chickens and send them to cover, or cluck like an old hen and bring 229 young chicks from their

---

[120] From "Birds of New England," by Wilson Flagg, used by special arrangement with the Page Co., Boston.

hiding-places. Some mockers seem to be able to reproduce the bird-songs they hear more melodiously than the singers themselves render them.

Mockingbirds' bravery in defense of their nests and their young is well known. They have an especial antipathy to dogs and cats, and are merciless in their attacks on those animals if seen near the vicinity of their nests. A friend in California told me that her cat was in abject terror of a mockingbird. Instead of considering him tempting prey, she invariably fled to cover when he appeared, and remained in hiding for a time. The fur on her sides was noticeably thinned where the angry bird had pulled out numerous locks. One day, while my family were visiting San Francisco, they heard a dog yelping piteously and discovered him running at lightning speed down the middle of the street. A mockingbird was perched on his back and was pulling hairs out of his tail with spiteful tweaks. Mockers have been known to kill snakes that approached their nests, and to attack human beings with great fury.

They like to live near people and seem to respond to the affection shown them in the South, where they are such favorites that they are seldom molested. Formerly mockingbirds were trapped for cage-birds, as were cardinals, but this practice is largely discontinued now, because of protective laws and aroused public sentiment.

Dr. Henry W. Henshaw says:

> "It is not surprising that the mockingbird should receive protection principally because of its ability as a songster and its preference for the vicinity of dwellings. Its place in the affections of the South is similar to that occupied by the robin in the North. It is well that this is true, for the bird appears 230 not to earn protection from a strictly economic standpoint. About half of its diet consists of fruit, and many cultivated varieties are attacked, such as oranges, grapes, figs, strawberries, blackberries, and raspberries. Somewhat less than a fourth of the food is animal matter, of which grasshoppers are the largest single element. The bird is fond of cottonworms, and is known to feed on the chinch bug, rice

weevil, and bollworm. It is unfortunate that it does not feed on injurious insects to an extent to offset its depredations on fruit."[121]

Professor Beal says, however, "The mockingbird will probably do little harm to cultivated fruits so long as wild varieties are accessible and abundant."[122] Wise cultivators of fruit take this into consideration and plant accordingly, to keep both their fruit and the delightful, amusing mockingbirds.

THE YELLOW-BILLED CUCKOO

Cuckoo Family—Cuculidæ

Length: About 12 inches; tail over 6 inches.

Male and Female: Brownish-gray above with a greenish tinge; white underneath; *reddish-brown wings*; feathers brightest on inner web; middle tail feathers brownish-gray; outer ones *black, broadly tipped with white, tips decreasing in size toward center*; *lower mandible of bill yellow* except at the end.

Notes: A rapid, guttural utterance of the words *cook-cook-cook-cook* and *cow-cow-cow-cow*. Our cuckoos sometimes give a cooing note, but do not say *cuck'-oo* like their European relatives.

Flight: Swift and difficult to observe, as the cuckoo glides rapidly from bough to bough, under cover if possible.

Nest: A loosely-constructed platform of sticks.

Habitat: Orchards, woodlands, park-like estates, and quiet shady streets. Cuckoos are occasionally seen in exposed, sunny places.

Range: Eastern North America. Breeds from southern Canada and northern United States as far west as North Dakota and as far south as northern Louisiana and Florida; winters south to Argentina.

---

[121] *Farmers' Bulletin 513*, U. S. Dept. of Agriculture, Biological Survey.

[122] From *Farmers' Bulletin 755*, U. S. Dept. of Agriculture, Biological Survey.

Figure 49. Yellow-billed cuckoo.

The black-billed cuckoo is similar to the Yellow-bill in general appearance, but has several marked differences. Its upper parts are more greenish; its tail-feathers have smaller white tips; its wings are *gray, not reddish-brown*; its bill is *black*, not *yellow*; its eye-ring is red.

Cuckoos seem to have less individuality than many of our birds; they resemble several of them. They are not unlike catbirds in their quiet, stealthy movements; they are slender, gray-and-white, and long-tailed like mockingbirds; they build nests somewhat like those of mourning doves.

They are shy, solitary birds, that are known by their note rather than by sight. I never heard of anyone but Wordsworth and Wilson Flagg who loved cuckoos or called them "darlings of the spring." The European cuckoo has, however, a very different nature and a more joyous note.

Burroughs is most amusing in his comments. He says:

> "We cannot hail our black-billed as 'blithe newcomer,' as Wordsworth does his cuckoo. 'Doleful newcomer,' would be a fitter title. There is nothing cheery or animated in his note, and he is about as much a 'wandering voice' as is the European bird. He does not babble of sunshine and of flowers. He is a prophet of the rain, and the country people call him the rain crow. All his notes are harsh and verge on the weird."[123]

He is, however, worthy of consideration. He is of great value to farmers and apple-growers because of his appetite for caterpillars and grasshoppers. Professor Beal wrote as follows:

> "The common observation that cuckoos feed largely on caterpillars has been confirmed by stomach examination. Furthermore, they appear to prefer the hairy and spiny species, which are supposed to be protected from the attacks of birds. The extent to which cuckoos eat hairy caterpillars is shown by the inner coatings of the stomachs, which frequently are so pierced by these hairs and spines that they are completely furred. The apple-tree tent-caterpillar and the red-humped apple-caterpillar are also eaten. In all, caterpillars constitute two-thirds of the total food of the yellow-billed cuckoo in the South. Few birds feed so exclusively upon any one order of insects.

---

[123] From "Under the Maples," by John Burroughs, pages 87 & 88.

"The natural food for cuckoos would seem to be bugs and caterpillars which feed upon leaves, as these birds live in the shade among the leaves of trees and bushes. Not so with grasshoppers, whose favorite haunts are on the ground in the blazing sunshine, yet these creatures are the second largest item in the cuckoo's diet. Grasshoppers are so agreeable an article of food that many a bird apparently forsakes its usual feeding grounds and takes to the earth for them. Thus it is with the cuckoos; they quit their cool, shady retreats in order to gratify their taste for these insects of the hot sunshine. But there are some members of the grasshopper order that live in the shade, as katydids, tree crickets, and ground crickets, and these are all used to vary the cuckoo's bill of fare."[124]

It eats, also, bugs that injure oranges and melons, and the cotton-boll weevil in large numbers.

THE LEAST FLYCATCHER OR CHEBEC

Flycatcher Family—Tyrannidæ

In March, there comes to us from the South the phœbe, inconspicuous in plumage, yet easy to identify because of its distinctive call. About a month later there arrives the smallest member of our Flycatchers,—the Chebec or Least-Flycatcher. Less than five and a half inches in length, slender, olive-brown above, grayish-white beneath with an indistinct grayish band across the breast, this little bird might escape our notice were it not for its oft repeated and unmistakable call-note. It announces its presence by uttering its name *Chebec*, as clearly and persistently as its cousins, the phœbe and pewee, say theirs.

The chebec is a bird to be found in orchards, by roadsides, and in trees of village streets. Like other members of its family it seeks conspicuous perches, from which it dives after flies, moths, and other insects, returning to its perch to wheeze out its name, with jerks and twitches of its tail.

---

[124] Farmers' Bulletin 755, Biological Survey, U. S. Dept. of Agriculture.

It breeds from central Canada to central United States as far south as Nebraska, Iowa, Indiana, Pennsylvania, and New Jersey, and in the Alleghany Mts. to North Carolina; winters from Mexico to Panama and Peru.

THE KINGBIRD

Flycatcher Family—Tyrannidæ

Length: About 8½ inches.

Male and Female: Upper parts dark gray; under parts pure white, with an indistinct grayish wash at the sides of the breast; *head grayish-black, slightly crested, with a concealed orange patch*; bill with bristles at the base; wing-feathers and upper tail-coverts tipped or edged with white; *tail fan-shaped in flight, showing a broad white band at the end.*

Note: An unmusical, rattling *Squeak-squeak? squeak-squeak-squeak?* uttered frequently, and apparently in an irritated mood. The sharply hooked beak and fierce-looking eye also give the appearance of pugnacity.

Habitat: Orchards, trees by roadsides, and near farm-buildings. One looks for the kingbird in open country, not in woodlands; he seeks conspicuous perches. The nests are placed in trees—in those of orchards preferably.

Range: North and South America. Breeds from South-central Canada and throughout the United States except in the south-west; winters from Mexico to South America.

No more interesting description of the Kingbird has come to my attention than that by Major Bendire. He writes as follows:

> "Few of our birds are better known throughout the United States than the Kingbird. Bold and fearless in character, yet tame and confiding in man, often preferring to live in close proximity to dwellings, in gardens
> 236 and orchards, they are prime favorites with the majority of our farming population, and they well deserve their fullest protection. Few

birds are more useful to the farmer; their reputation for pugnacity and reckless courage is so well established that it is almost needless to dwell on it, as it is well known that they will boldly attack and drive off the largest of our Raptores, should one venture too near to their chosen nesting-sites.

"Where a pair or more of these birds make their home in the vicinity of a farmhouse, the poultry yard is not likely to suffer much through feathered marauders at least; they are a perfect terror to all hawks, instantly darting at them and rising above them, alighting on their shoulders or necks, and picking away at them most unmercifully until they are only too willing to beat a hasty retreat. The male is seemingly always on the lookout from his perch on the top branches of a tree or post for such enemies and no matter how large they may be, a pair of Kingbirds is more than a match for any of them, our larger Falcons and Eagles not excepted. Crows and Blue Jays seem to be especially obnoxious to them, and instances are on record where they have done them material injury."

Major Bendire says also that kingbirds do not "bully" all birds, but "as a rule live in harmony with them, protecting not only their own nests but those of their small neighbors as well, who frequently place their nests within a few feet of the Kingbirds—the Orchard Oriole, for instance." He tells however, of the kingbird's dislike of the hummingbird—that he has twice seen the tiny "aggressor" put the larger bird to flight.[125]

Kingbirds were for a long time believed to eat bees and therefore were in disfavor. They were called Bee-birds or Bee-Martins and were shot by bee-keepers who did not understand their great value. Professor Beal and other investigators in the Biological Department at Washington have discovered that ninety per cent of kingbirds' food consists of insects, mostly injurious beetles that prey upon grain and fruit. They occasionally eat bees, but examination of many stomachs reveals a marked preference for drones over workers, and for wasps, wild bees, and ants over hive bees.[126] So kingbirds have been exonerated.

---

[125] From "Life Histories of North American Birds," by Major Charles Bendire.

[126] *Farmers' Bulletin 630*, Biological Survey, U. S. Dept. of Agriculture.

Figure 50. Kingbird.

The Gray Kingbird

Flycatcher Family—Tyrannidæ

Length: About 9 inches.

Male and Female: Upper parts light gray, darker about the cheeks; concealed orange patch on the crown; under parts whitish, washed with gray on the breast; wings and tail brownish; *no white band on the tail*, like the northern kingbird; *bill very heavy—almost an inch long*, with bristles at the base.

Note: A loud call, *Pit-tear'-re*, "which is constant and is at times lengthened and softened until it might almost be called a song."[127] The natives of Porto Rico call the bird "pitir're" because of its note.

Range: Breeds from Georgia, southeastern South Carolina, Florida, and Yucatan, through the Bahamas and West Indies to northern South America; winters from the Greater Antilles southward. It is common in our southeastern states.

The following is an extract from Dr. Wetmore's interesting description of the Gray Kingbird in the bulletin, "Birds of Porto Rico," used with the permission of the author:

> "The gray kingbird has the reputation among the country people of being the earliest riser among birds. In the daytime it scatters along the slopes and through the fields to feed, but at nightfall gathers in small parties along streams to roost in the bamboos or in the mangroves surrounding the lagoons. The nesting season extends from April to July and during the latter month young are abundant. At all times very pugnacious, pursuing blackbirds, hawks, and other birds, they now become doubly so, resenting all intrusions in their neighborhood. Occasionally they were seen standing on open perches during showers with outspread trembling wings, evidently enjoying the downpour.
>
> "A few facts regarding the insect food of this kingbird were learned from field observation. Birds were twice observed eating the caterpillars of a large sphinx moth. These were beaten on a limb, and then the juices were extracted by working the body through the bill, while only the skin was discarded. Their services in eating these and other caterpillars were recognized."

---

[127] Biological Survey Bulletin, No. 326, "Birds of Porto Rico," by Alex. Wetmore.

The Crested Flycatcher

Flycatcher Family—Tyrannidæ

Length: About 9 inches.

Male and Female: *Olive-gray above; throat and breast light gray; belly, bright yellow; head conspicuously crested*; bill, long, dark, slightly hooked, with bristles at its base; wings brown, margined with white, pale yellow, and reddish-brown; middle tail-feathers, dull brown; inner web of other tail-feathers reddish-brown.

Notes: A whistle that attracts attention. Major Bendire describes the "Great Crest's" notes as follows:

"It utters a variety of sounds; the most common is a clear whistle like *e-whuit-huit*, or *wit-whit, wit-whit*, repeated five or six times in a somewhat lower key, and varied to *whuir*, *whuree*, or *puree*, accompanied by various turnings and twistings of the head. Its alarm-note is a penetrating and far-reaching *whēēk, whēēk*."

Nest: The nest of the crested flycatcher is unique. Major Bendire says that it "is usually placed in a natural cavity of some tree or dead stump; possibly in an abandoned woodpecker excavation, though a natural one is preferred." He says also that "nests vary in bulk; are begun with a base of coarse trash and finished with fine twigs, bunches of cattle hair, pine needles, dry leaves and grasses, the tail of a rabbit, pieces of catbirds' eggshells, exuviæ of snakes, owl and hawk feathers, tufts of woodchucks' hair and fine grass roots."

Snake-skins "seem to be present in the majority of 240 the nests of this species; sometimes in the nest proper, and again placed around the sides of it, in all probability for protective purposes, and changed and rearranged from time to time" ... probably hung outside to "alarm intruders."[128]

The Crested Flycatcher lives in eastern North America; breeds from southern Canada to Florida, and winters in Mexico and northern South America. He is a common summer resident of the Middle and Southern

[128] From "Life Histories of North American Birds," by Major Charles Bendire.

States especially. Though louder-voiced than his relatives, the kingbird, phœbe, and wood pewee, he is not so well known because he is shyer. He is not so pugnacious as the kingbird, but he is known to fight fiercely for a mate.

The Olive-Sided Flycatcher

Flycatcher Family—Tyrannidæ

Length: About 7½ inches.

Male and Female: Upper parts *and sides olive-gray*, the gray extending across the breast; throat and belly yellowish, the yellow extending in a point almost to the center of the breast; a patch of whitish feathers on both sides of the back near the rump; head slightly crested; bill long, black above, yellow below, bristles at the base, wings and tail olive-brown.

Notes: A monotonous call-note, *Pit-pit-pit*, and a loud, clear *Peep-here* or *Peep-peep-here*, frequently uttered from the top of a tall spruce.

Habitat: Groves of conifers.

Range: North and South America. Breeds from central 241 Alaska and Canada, in coniferous forests of western United States to northern Michigan, New York, and Maine, south to the mountains of North Carolina; winters in South America from Colombia to Peru.

The Wood Pewee

Flycatcher Family—Tyrannidæ

Length: From 6 to 6½ inches.

Male and Female: Dark olive-gray above, darkest on the head, which is somewhat crested; the slightly hooked bill has bristles at its base; under parts, grayish-white, slightly tinged with yellow; breast and sides a darker gray; brownish wings and tail; two whitish wing-bars that are more conspicuous than those of the phœbe.

Notes: *Pee-a-wee*, uttered slowly and mournfully, yet with sweetness and tenderness. Sometimes the phrase is followed by an abrupt *Peer*, given with a falling inflection. At times pewees sing continuously. My sister timed one that sang for an hour and twenty minutes at daybreak.

Nest: One of the most beautiful made. It is rather broad and flat, decorated on the outside with lichens similarly to that of the hummingbird. The nest seems to grow out of the branch on which it is placed.

Range: North and South America. Breeds from southern Canada to southern Texas and central Florida, westward to eastern Nebraska; winters from Nicaraugua to Colombia and Peru.

Of all the flycatchers of my acquaintance the Wood Pewee is the most lovable. He is the only one that possesses a sweet voice; but his note, long-drawn and sad, seems to proceed from an over-burdened heart. The appearance of the little bird is dejected, as with drooping tail, he utters the plaintive sound.

The nature of the pewee is sweet and trustful. I have always found him responsive, replying almost invariably as I have imitated his note. I once had a particularly pleasant experience and succeeded in convincing a little pewee of my friendly attitude toward him. One summer I was obliged to spend many weary days in a hammock hung in a grove; I beguiled the tedious hours by endeavoring to attract birds to close proximity. A pewee came oftenest; he frequently perched on a bough within a few feet of my hammock, and “talked back” to me between dives after insects. That he knew me and was unafraid was proved, for when relatives and friends arrived later in the summer, he would fly away at their approach.

I saw much of him, even when parental responsibilities claimed him. One day, after the young had flown, I came upon him calling earnestly, evidently to a fledgling that was on the ground at my feet. I picked up the little thing; it cuddled down in my warm hand and closed its eyes. Its father continued to call, but without excitement at such a proceeding; he seemed to know that I would not hurt his baby. I put it on a bough near him and left them to work out their bird-problems together.

Figure 51. Wood pewee.

Not many days later, we saw four young pewees perched in a row on a wire near the house, with their parents in attendance. The father called repeatedly and the little ones made sweet inarticulate gurglings, finding their voices. They were as dear a bird-family as it has ever been my pleasure to see.

Dallas Lore Sharp, in his delightful essay, "A Palace in a Pig-pen," thus summarizes the flycatchers:

> "Not much can be said of this flycatcher family, except 244 that it is useful—a kind of virtue that gets its chief reward in heaven. I am acquainted with only four of the other nine eastern members, [besides the phœbe], the great crested flycatcher, kingbird, wood pewee, and chebec,—and each of these has some redeeming attribute besides the habit of catching flies.
>
> "They are all good nest-builders, good parents, and brave, independent birds; but aside from phœbe and pewee—the latter in his small way the sweetest voice of the oak woods—the whole family is an odd lot, cross-grained, cross-looking, and about as musical as a family of ducks. A duck seems to know that he cannot sing. A flycatcher knows nothing of his shortcomings. He believes that he can sing, and in time he will prove it. If desire and effort count for anything, he certainly must prove it in time. How long the family has already been training no one knows. Everybody knows, however, the success each flycatcher of them has thus far attained. It would make a good minstrel show, doubtless, if the family would appear together. In chorus, surely, they would be far from a tuneful choir. Yet individually, in the wide universal chorus of the out-of-doors, how much we should miss the kingbird's metallic twitter and the chebec's insistent call!"[129]

The Ruby-Crowned Kinglet

Old World Warbler Family—Sylviidæ

Length: A little over 4¼ inches.

Male: Olive-green above, buff underneath, *a ruby-red crown*; wings brown, edged with olive-green; two light wing-bars; tail brown, forked.

Female: Similar to male, but lacking the red crown. The females resemble tiny warblers in appearance.

---

[129] From "The Whole Year Round," by Dallas Lore Sharp.

Note: A sharp scolding-note.

Song: A wonderful song,—full, loud, and indescribably beautiful. It is hard to believe that so finished and remarkable a song could come from so small a bird.

Habitat: Woods, thickets, and orchards. Kinglets are usually seen near the ends of branches.

Range: Northern North America. Breeds in the tree-regions of southern Canada, southern Alaska, and the higher mountains of the western United States. Like many of the warblers, the Ruby-crowned Kinglet is a spring and fall migrant, and its arrival is therefore of especial interest. It excels most of the warblers in its power of song, and is even more agile than they.

In Bulletin 513 of the Biological Survey is the following description of the Ruby-crown: "In habits and haunts this tiny sprite resembles a chickadee. It is an active, nervous little creature, flitting hither and yon in search of food, and in spring stopping only long enough to utter its beautiful song, surprisingly loud for the size of the musician. Three-fourths of its food consists of 246 wasps, bugs, and flies. Beetles are the only other item of importance. The bugs eaten by the kinglet are mostly small, but, happily, they are the most harmful kinds. Treehoppers, leafhoppers, and jumping plant-lice are pests and often do great harm to trees and smaller plants, while plant-lice and scale insects are the worst scourges of the fruit-grower—in fact, the prevalence of the latter has almost risen to the magnitude of a national peril. It is these small and seemingly insignificant birds that most successfully attack and hold in check these insidious foes of horticulture. The vegetable food consists of seeds of poison ivy, or poison oak, a few weed seeds, and a few small fruits, mostly elderberries."

THE BLUE-GRAY GNATCATCHER

Old World Warbler Family—Sylviidæ

Length: About 4½ inches.

Male: Bluish-gray above; grayish white below; forehead black, black line over the eye; slender, curving bill; wings dark gray, edged with grayish-white; tail long, outer tail-feathers nearly all white; middle tail-feathers black; tail elevated and lowered frequently.

Female: Similar to male, but without the black forehead; line over eye indistinct.

Call-note: A nasal *tang*.

Song: A delightful song,—sweet, but not strong.

Habitat: Woodlands, where it usually frequents treetops.

Range: Southeastern United States. Breeds from eastern Nebraska, southern Wisconsin, Michigan, and Ontario, southwestern Pennsylvania, Maryland, and southern New Jersey to southern Texas and central Florida; 247 winters from northern Florida to the West Indies and central America; casual in Minnesota, New England, and New York.

This dainty little sprite partakes of the qualities of a number of birds. Like the warblers, it is insectivorous and inhabits treetops; like its relative, the ruby-crowned kinglet, it has a finished and wonderful song; like the wrens it has a habit of cocking its tail nervously; while its long black and white tail reminds one of the mockingbird. It is an especially pretty sight, fluttering about the moss-hung trees of Florida.

The Red-Eyed Vireo

Vireo Family—Vireonidæ

Length: About 6¼ inches.

Male and Female: Olive-green above, silvery white below; *crown gray, bordered with a narrow black line; a broader white line over the eye, a dark streak through the eye; iris red* or *reddish-brown*; wings and tail grayish-green, edged with olive.

Habitat: In open woodlands and along well-shaded roads.

Range: North and South America. Breeds from central Canada, northwestern, central, and eastern United States, to central Florida; winters in South America.

Note: A nasal *whăh*, that sounds ill-natured and unpleasant.

Song: A series of phrases—incessant, monotonous,—that continue from morning until night, and during August, when most birds are quiet. Wilson Flagg called the Red-eye the "Preacher-bird" and wrote of him as follows:

> "The Preacher is more generally known by his note, because he is incessant in his song, and particularly vocal during the heat of our long summer days, when only a few birds are singing. His style of preaching is not declamation. Though constantly talking, he takes the part of a deliberative orator, who explains his subject in a few words and then makes a pause for his hearers to reflect upon it. We might suppose him to be repeating moderately, with a pause between each sentence, 'You see it—you know it—do you hear me?—do you believe it?' All these strains are delivered with a rising inflection at the close, and with a pause, as if waiting for an answer.

"He is never fervent, rapid, or fluent, but like a true zealot, he is apt to be tiresome from the long continuance of his discourse. When nearly all other birds have become silent, the little preacher still continues his earnest harangue, and is sure of an audience at this late period, when he has few rivals."[130]

Mr. Forbush discovered that this preacher "practiced as he preached," and tells us of his own observations in the following words:

> "One sunny day in early boyhood I watched a vireo singing in a swampy thicket. He sang a few notes, his head turning meanwhile from side to side, his eyes scanning closely the nearby foliage. Suddenly his song ceased; he leaned forward,—sprang to another twig, snatched a green caterpillar from the under side of a leaf, swallowed it, and resumed his song. Every important pause in his dissertation signalized the capture of a larva. As the discourse was punctuated, a worm was punctured. It seems as if the preaching were a serious business with the bird; but this seeming is deceptive, for the song merely masks the constant vigilance

[130] From Wilson Flagg's "Birds of New England," used with permission of The Page Co., Boston.

and the sleepless eye of this premium caterpillar-hunter. In the discovery of this kind of game the bird has few superiors."[131]

Figure 52. Red-eyed vireo.

This vireo builds a very attractive nest of strips of bark and fiber, a soft basket hung at the fork of a branch. I recall one nest suspended only a few feet from the ground in a low tree on Cape Cod. We came upon the nest so

[131] From "Useful Birds and Their Protection," by E. H. Forbush, p. 205.

suddenly that the little brooding mother looked 250 at us with frightened eyes, but she remained at her post, and soon learned that we meant no harm. Many times a day we went by her precious cradle. At night we passed quietly, so as not to waken the faithful little mother-bird with her head tucked under her wing. Our flashlight never once disturbed her. Mr. Forbush says, "This vireo sleeps very soundly, and is sometimes so oblivious to the world that she may be approached and taken in the hand."[132]

Burroughs wrote:

> "Who does not feel a thrill of pleasure when, in sauntering through the woods, his hat just brushes a vireo's nest?... The nest was like a natural growth, hanging there like a fairy basket in the fork of a beech twig, woven of dry, delicate, papery, brown and gray wood products,—a part of the shadows and the green and brown solitude. The weaver had bent down one of the green leaves and made it a part of the nest; it was like the stroke of a great artist. Then the dabs of white here and there, given by the fragments of spider's cocoons—all helped to blend it with the flickering light and shade."[133]

THE WARBLING VIREO

Vireo Family—Vireonidæ

Length: About 5¾ inches.

Male and Female: Grayish-olive above; *indistinct whitish line over eye*; under parts grayish-white with a faint yellowish tinge; *no bars on wings; iris dark brown*, not reddish.

Note: A nasal *yăh*, not unlike the call-note of the red-eyed vireo.

Song: A sweet continuous warble, with a rising inflection at the end. It sounds like a whistled *Whew-whew-whew whew-whew-whew-whee?*

---

[132] From "Useful Birds and Their Protection," by E. H. Forbush, p. 205.

[133] From "Under the Maples," by John Burroughs, p. 99.

Habitat: Parks and shaded village streets. Its neutral coloring and its preference for treetops make it difficult to distinguish. Its cheerful, pleasant song is the surest means of identification.

Range: Eastern North America. Breeds from south-central Canada to northwestern Texas, southern Louisiana, North Carolina, and Virginia; winters south of the United States, though exact locality is unknown. Not nearly so widely distributed as the red-eyed vireo.

This vireo, like other members of its family, is an indefatigable devourer of insects. Mr. Forbush reports that it feeds on flies, mosquitoes, and grasshoppers, but that its chief food consists of caterpillars and other leaf-eating insects, especially the elm-leaf beetle; consequently it is found frequently in elm-shaded streets and yards.

THE WHITE-EYED VIREO

Vireo Family—Vireonidæ

The White-Eyed Vireo differs from his red-eyed cousin in being slightly smaller, in having a *small patch of yellow around the eye*, a *white iris*, and *two wing-bars*. His head is greener and his breast and sides are tinged with yellow.

He lives in thickets. He possesses in a marked degree the vireo habit of scolding. He has more power as a songster than his better-known relatives. Mr. Chapman 252 describes him most delightfully as follows: "If birds are ever impertinent, I believe this term might with truth be applied to that most original, independent dweller in thickety under-growths, the white-eyed vireo. Both his voice and manner say that he doesn't in the least care what you think of him; and, if attracted by his peculiar notes or actions, you pause near his haunts, he jerks out an abrupt 'Who are you, eh?' in a way which plainly indicates that your presence can be dispensed with. If this hint is insufficient, he follows it by a harsh scolding, and one can fancy that in his singular white eye there is an unmistakable gleam of disapproval.

"I have always regretted that the manners of this Vireo have been a bar to our better acquaintance, for he is a bird of marked character and with unusual vocal talents. He is a capital mimic, and in the retirement of his home sometimes amuses himself by combining the songs of other birds in an intricate pot-pourri."[134]

THE YELLOW-THROATED VIREO

Vireo Family—Vireonidæ

The Yellow-throated Vireo resembles the White-eye in being olive-green above, yellowish underneath, and in having two distinct white wing-bars. He differs in possessing a *bright yellow throat*, *breast*, and *ring* about a *dark eye*.

Mr. Forbush says of this bird; "The song is a little louder than that of most vireos, and may be easily distinguished 253 from all others. It usually consists of two or three rich and virile notes, uttered interrogatively or tentatively, followed immediately by a few similar tones uttered decisively. The bird appears to ask a question, and then answer it. Its alarm notes are as harsh as those of an oriole, and somewhat similar in quality."[135]

THE WARBLERS

Warbler Family—Mniotiltidæ

No family of birds is more difficult for a beginner to identify than the warblers. Reasons for this fact are various. In the first place, warblers are small and agile, and usually inhabit treetops, where it is hard to see their plumage. The number of the species is large,—155 species are known, 74

[134] From "Birds of Eastern North America," by Frank M. Chapman.
[135] From "Useful Birds and Their Protection," by E. H. Forbush, p. 208.

of which are found in North America, and 55 in the United States alone. Some of the males wear a "Joseph's coat of many colors"; some of the females are so different from their mates as to puzzle an observer, and the young birds frequently differ from both parents. Then, too, most warblers are not gifted songsters, but utter only a weak trill. A number of them are seen only during their migration to northern woods; they linger too short a time to become more than passing bird-acquaintances.

Warblers are insectivorous and do not arrive until the earth teems with insect life. Most of them depart for the South as soon as insects begin to decrease in number or disappear. They are very shy and migrate at night.

Many are the disasters that befall them when they journey near the sea-coast. In Dr. Wells W. Cooke's article entitled "Our Greatest Travelers" are the following statements: "It is not to be supposed that these long flights over the waters can occur without many casualties, and not the smallest of the perils arises from the beacons 255 which man has erected along the coast to insure his own safety. 'Last night I could have filled a mail-sack with the bodies of little warblers which killed themselves striking against my light,' wrote the keeper of Fowey Rocks lighthouse, in southern Florida.

"Nor was this an unusual tragedy. Every spring the lights along the coast lure to destruction myriads of birds who are en route from their winter homes in the South to their summer nesting-places in the North. Every fall a still greater death-toll is exacted when the return journey is made. A red light or a rapidly flashing one repels the birds, but a steady white light piercing the fog proves irresistible."[136]

Few people realize the great good done by warblers. Mr. Forbush says that in migration they seem to possess enormous appetites. A Hooded Warbler was found to catch on the average two insects a minute or one hundred and twenty an hour. At this rate the bird would kill at least nine hundred and sixty insects a day, in an eight hour working day!

Dr. Judd reported a Palm Warbler that ate from forty to sixty insects a minute. In the four hours he was under observation he must have eaten

[136] From "Our Greatest Travelers," by Wells W. Cooke, of the Biological Survey.

nine thousand, five hundred insects. Mr. Forbush says that he has seen warblers eating from masses of small insects at such a rate that it was impossible for him to count them.[137]

Identification of Warblers

In order to identify warblers, most people need to group them in some way. The following grouping of my own has helped me to recognize and remember the more common species:

The Ground Warblers

1. The Ovenbird
2. The Water Thrushes
3. The Worm-eating Warbler
4. The Palm Warblers

Black and White Warblers

1. The Black and White Creeping Warbler
2. The Black-poll Warbler

Black, White, and Yellow Warblers

1. The Myrtle or Yellow-rumped Warbler
2. The Magnolia Warbler

Black and Orange Warblers

1. The Redstart
2. The Blackburnian Warbler

---

[137] From "Useful Birds and Their Protection," by E. H. Forbush, pages 185 and 186.

Warblers with Yellow or Olive-green Predominating

1. The Yellow Warbler
2. The Pine Warbler
3. The Maryland Yellow-throat
4. The Hooded Warbler
5. Wilson's Warbler
6. The Black-throated Green Warbler
7. The Canadian Warbler
8. The Yellow-breasted Chat
9. The Yellow Palm Warbler

Warblers with Blue or Blue and Yellow Predominating

1. The Cerulean Warbler
2. The Black-throated Blue Warbler
3. The Blue-winged Warbler
4. The Golden-winged Warbler
5. The Parula Warbler

Warblers with Reddish-brown Markings

1. The Bay-breasted Warbler
2. The Chestnut-sided Warbler

Group One—The Ground Warblers

The Oven-Bird

Length: A little over 6 inches.

Male and Female: Olive-brown above; head with a golden-brown crown, bordered with two black lines that extend from bill to neck; under parts white; a brown streak at each side of the throat; breast and sides heavily streaked with black; no bars on wings, or patches on tail.

Note: Mr. Forbush interprets the oven-bird's note as "*chick,*' kerchick,' kerchick,' repeating the phrase an indefinite number of times."[138] John Burroughs has rendered it as, "teacher, *teacher*, teacher, TEACHER, *TEACHER*." The bird is frequently spoken of as the "Teacher-bird."

Song: A "flight song" which Mr. Forbush describes as follows: "When I lingered in the woods at evening until the stars came out, I heard a burst of melody far above the treetops, and saw the little singer rising against the western sky, simulating the Skylark, and pouring forth its melody, not to the orb of day but to the slowly rising moon; then, when the melody came nearer, the exhausted singer fell from out the sky and shot swiftly downward, alighting at my very feet."[138]

Habitat: Woodlands, where the oven-bird spends much of its time on the ground.

Range: North America. Breeds in the forests of Canada and the United States to Kansas, southern Missouri, Ohio Valley, Virginia, and in the mountains of Georgia and South Carolina; winters from central Florida to Colombia.

This shy forest-dweller is little seen except by the tireless haunter of woods. I well remember my first quest for the owner of a voice that seemed to proceed from every part of the small grove I was searching. His ventriloquistic power led me on until I was about to give up in weariness and discouragement, when suddenly I came upon this golden-crowned warbler that had made the woods ring. He seemed very small for so loud a vocalization.

Another day, quite by accident, I discovered his oven-shaped nest:

> "Arched and framed with last year's oak-leaves,
> Roofed and walled against the raindrops."[139]

---

[138] From "Useful Birds and Their Protection," by E. H. Forbush.
[139] From "The Oven-Bird," by Frank Bolles.

Since that time I have had numerous views of oven-birds. One in particular, seemed quite unafraid; and several times approached within a few feet of where I was seated.

Figure 53. Oven-bird.

The Water-Thrushes

Water-Thrushes: The Water-thrushes resemble the oven-bird in size and general appearance. Their crowns are dark instead of golden; the northern water-thrush has a *light* line over the eye, and a bright yellow streaked breast; the Louisiana water-thrush a conspicuous *white* line over the eye, buff sides, and white under parts.

Both birds, as their name implies, love the vicinity of forest brooks. Both walk instead of hop, and "*tip-up*" when they alight. They are wonderful songsters, but are not widely known.

Range: Eastern North America. The Northern Water-thrush breeds in east-central Canada, northwestern New York, northern New England, and in mountains south to West Virginia; winters in the West Indies and from the valley of Mexico to British Guiana.

The Louisiana Water-thrush is found from the northern parts of the United States south to Texas, Georgia, and South Carolina; winters from Mexico to Colombia.

The Worm-Eating Warbler

Length: About 5½ inches.

Male and Female: Back, wings, and tail olive-green, without white markings; head with two narrow and two broad black stripes, alternating with three cream-colored stripes; under parts cream-colored, lighter on throat and belly.

Song: A weak trill.

Habitat: "The Worm-eating warbler seems to prefer dense undergrowth in swampy thickets and wet places, ... wooded hillsides and ravines, and dense undergrowths of woodland.... The nesting site is on the ground."[140]

Range: Eastern North America from southern Iowa, northern Illinois, western Pennsylvania, and the lower Hudson and Connecticut valleys, south to Missouri, Tennessee, Virginia, and the mountains of South Carolina.

---

[140] From Eaton's "Birds of New York," page 383.

Group Two—The Black and White Warblers

The Black-and-White Warbler

Length: About 5¼ inches.

Male: Black, streaked with white—*no yellow*; head with broad black and white stripes; body with narrow stripes; white stripe over eye, black patch back of eye; striped throat and sides, white belly; tail grayish black; outer tail-feathers with white patches on inner web; wings black, with two distinct white bars.

Female: Similar, but with gray cheeks and whiter under parts, fainter streaks, and brownish sides.

Song: A thin, unmusical *se-se'-se-se'-se-se'-se-se'*.

This Black-and-White Warbler is as easy to identify as a zebra, because of its conspicuous black and white stripes. As it is found on tree-trunks, it is sometimes confused with the brown creeper. Its bill, however, is not curved like the creeper's, nor is its tail used as a prop. It resembles the nuthatch in its ability to *descend* as well as *ascend* tree-trunks.

These warblers, though they obtain their food from trees, nest on the ground in nests not unlike those of the oven-bird.

The Black-Poll Warbler

Length: About 5 inches.

Male: A black crown and white cheeks, giving the effect of a black cap pulled down over the eyes; throat and belly white; back and sides gray, streaked with black; two white wing-bars; two outer tail-feathers with white spot near tip.

Female: Olive-green above, streaked with black; breast and sides with yellowish wash.

Range: Widely distributed; common in the East during migration. Breeds in the forests of Alaska and north-central Canada, in Michigan, northern Maine, and the mountains of New Hampshire and Vermont.

Black-poll Warblers are similar in coloring to the black-and-white warblers, but are duller and less striking in appearance. In the breeding season, father, mother, and young differ in plumage, though a practiced eye may see resemblances, but in the fall they don coats so similar that they seem to have adopted a family costume.

The migration of black-poll warblers is interesting. Dr. Wells W. Cooke says:

> "All black-poll warblers winter in South America. Those that are to nest in Alaska strike straight across the Caribbean Sea to Florida and go northwestward to the Mississippi River. Then the direction changes and a course is laid almost due north to northern Minnesota, in order to avoid the treeless plains of North Dakota. But when the forests of the Saskatchewan are reached, the northwestern course is resumed, and, with a slight verging toward the west, is held until the nesting site in Alaska spruces is attained."[141]

Group Three—The Black, White, and Yellow Warblers

The Myrtle Warbler or Yellow-Rumped Warbler

Length: A little over 5½ inches, one of the larger warblers.

Male: The grayish upper parts, white under parts, (both streaked with black), and the black cheeks of the Myrtle Warbler remind one of the Black and White Creeping Warbler. Its *four patches of yellow*,—on the *crown*, *rump*, and on *each side* are distinctive. The wings and tail are

[141] "Our Greatest Travelers," by Wells W. Cooke, of the Biological Survey.

brownish-gray; wings, with *two white bars*; tail with graduated patches of white *near end* of outside feathers; white throat and belly.

Female: Browner above; breast less heavily streaked with black.

Notes: The notes and song of this warbler are described by Mr. Forbush as follows:

> "The Myrtle Warbler has a variety of notes, but the one usually uttered spring and fall is a soft chirp or *chup*, which, at a little distance, exactly resembles the sound produced by a large drop of water as it strikes the ground or leaf-mold. These sounds are so similar that after storms in the woods I have often found it difficult to distinguish the note of this warbler from the splash of the large drops that were still falling from the trees. The song is a rather weak warble, very sweet, and often of long duration.... It has quite as many variations as the song of any warbler that I now recall."[142]

Range: Breeds in the forest-belt of Canada and Alaska, south to Minnesota, Michigan, New Hampshire, Maine, 263 Vermont, Massachusetts, and the mountains of New York, winters from Kansas, New Jersey, southern New England to West Indies, Mexico, and Panama, and from central Oregon to southern California.

The Myrtle or Yellow-rumped Warbler is found in North America except in the western United States. It is so abundant and so distinctly marked as to be better known than many warblers.

> "Trim of form and graceful of motion, when seeking its food it combines the methods of the wrens, creepers, and flycatchers. This bird is so small and nimble that it successfully attacks insects too minute to be prey for larger birds. Flies are the largest item of food; in fact only a few flycatchers and swallows eat as many flies as this bird."[143]

---

142 "Useful Birds and Their Protection," E. H. Forbush, page 202.
143 Farmers' Bulletin, Biological Survey, Henry W. Henshaw.

The Myrtle Warbler is especially fond of bayberries and may be found, even in winter, where these berries are to be obtained. New Jersey and Cape Cod are favorite feeding places.

The Magnolia Warbler

Length: About 5 inches.

Male: Smaller than the Myrtle Warbler, and at first glance, not unlike it in appearance, because each bird has a yellow rump, a striped breast, dark gray upper parts, and back and breast streaked with black. The head of the Magnolia Warbler, however, has *no yellow patch*, but a *broad white line over the eye*, black cheeks and forehead, and *yellow* under parts, (instead of white), which are heavily streaked with black. The wings have *large white patches* instead of *bars*; the tail is black, with a broad white band extending *across* the *middle*,—a distinguishing mark.

Female: Similar to male, but duller.

Song:

> "It is one of our full-voiced warblers, the song resembling the syllables wee-to, *wee-to, wee-a-tee or witchi, witchi, witchi, tit, witchi-tit, witchi-tit, witchi-tit*, the first four words deliberate and even, the last three hurried and higher pitched.... The song is louder than the yellow warbler's."[144]

Habitat:

> "Throughout the migration season, the Magnolia warbler is common throughout our orchards and shade trees, as well as woodlands.... In its nesting grounds, this warbler prefers coniferous growth, especially young spruces."[144]

---

[144] From Eaton's "Birds of New York," pages 408, 409, 410.

Range: Breeds from southern Mackenzie, Keewatin, northern Quebec, and Newfoundland to central Alberta, Saskatchewan, Minnesota, northern Michigan, and northern Massachusetts; in the mountains of West Virginia, Maryland, Pennsylvania, New York; winters from southern Mexico to Panama.

Mr. C. F. Stone in "Birds of New York" says:

> "Every hemlock-clad gully or hemlock woods where the trees are close and limbs intertwined afford suitable haunts for this lively and emphatic singer.... Among the smaller gullies 1 or 2 pairs may be found, and in the larger gullies it is not unusual to locate 12 or 15 pairs during the nesting period. In some of these situations the Magnolia does not seem to occur, perhaps because it is so persecuted by red squirrels and cowbirds. The latter seems to make a specialty of presenting this warbler with one or more of its eggs, generally puncturing the eggs of the Magnolia before leaving the nest."[144]

Group Four—The Black and Orange Warblers

The Redstart

Length: About 5½ inches.

Male: Body glossy black, with a white belly, orange patches at the sides of the body and under the wings; an orange band across the wings; middle tail-feathers black; other tail-feathers broadly tipped with black but largely orange, conspicuous in flight; bill with bristles.

Female: Gray and olive-green above, white underneath; yellow instead of orange on sides, wings, tail, and under tail.

Young Male: Like female till end of first breeding season.

Nest: A beautiful structure made of strips of bark, root-fibers, and plant-down, and placed in the fork of a tree. If built in a birch sapling and decorated with bits of birch bark, it seems a part of the tree.

Song: A cheerful trill, rather weak and unmusical.

Range: North America. Breeds from Oklahoma, Arkansas, and North Carolina northward; winters in the West Indies, central Mexico, and northern South America.

The Redstart is one of the most beautiful and conspicuous of the warblers. Its fan-shaped, flame-colored tail tipped with black is its most distinctive mark. It is in almost constant motion, fluttering incessantly in pursuit of its insect prey. Mr. Forbush writes,

> "In all its movements its wings are held in readiness for instant flight, and in its sinuous twistings and turnings, risings and fallings, its colors expand, contract, and glow amid the sylvan shades like a dancing torch."[145]
>
> Like flycatchers, the redstart has bristles at the base of its bill, which makes the capture of a great variety of insects an easy matter. It has been named the "flycatcher of the inner treetops, but it is a flycatcher of the bushtops as well."[146]

The Blackburnian Warbler

Length: A little over 5 inches.

Male: Black crown, with bright orange patch in the center; irregular black patch extending from eye, bordered with orange; throat and breast orange, becoming yellowish on belly; back black, streaked with white; sides streaked with black; wings black, with white edges and a *large white patch*; tail black, *most of the feathers nearly all white on inner web*. Colors duller in the fall.

Female: Upper parts grayish-olive, streaked with white; orange parts paler, less white on wings and tail.

Song: A "thin" warbler-like trill.

Habitat: Treetops of coniferous forests preferably.

---

[145] "Useful Birds and Their Protection," by E. H. Forbush, pages 196 and 198.

[146] "Useful Birds and Their Protection," by E. H. Forbush, pages 196 and 198.

Range: Breeds from central Canada to northern United States, and in the Alleghany Mts. from Pennsylvania to Georgia; winters in Colombia and Peru.

This brilliant warbler flashes flame as do the oriole and the redstart, and like them, always brings a thrill of pleasure. It remains with us so short a time that its appearance is an event.

Mr. Forbush tells of going out at daybreak May 11, 1900, at Amesbury, Mass., to observe the migrant warblers. He says:

> "As we walked through the streets 267 of the village, many male Blackburnian Warblers were seen among the street trees. A little later we saw them in the orchards, their brilliant orange breasts flashing in the sunlight. As we approached the woods it was everywhere the same. The night had been very cold, and other insect-eating birds were seeking benumbed insects on or near the ground. There were four bright Redstarts flitting about on the upturned sod of a newly plowed garden. These and other species of Warblers were to be seen in every orchard, wood, and thicket. The Blackburnian Warblers had come in during the night, and were busy hunting for their breakfasts until 7 o'clock, when we went to ours. At 8 o'clock not a single Blackburnian was to be seen. I scoured the country till nearly noon, finding all the other Warblers as at daybreak, but not a Blackburnian could be found. They had done their share of the good work and passed on. A later riser would have missed them."[147]

Eaton says:

> "The Blackburnian warbler during the migration season associates with the Magnolia, Bay-breasted, and Chestnut-sided warblers among the blossoming fruit trees and the leaving shrubbery and shade trees of our lawns and parks. During the nesting season, however, it is almost entirely confined to mixed and evergreen forests, being especially fond of hemlocks and spruces.... The old name of Hemlock warbler is perfectly appropriate. The Blackburnian flutters about while feeding almost as

[147] From "Useful Birds and Their Protection," by E. H. Forbush, page 102.

conspicuously as the Redstart and Magnolia, displaying its brilliant colors and pied pattern very effectively."[148]

GROUP FIVE—THE YELLOW AND OLIVE-GREEN WARBLERS

The Yellow Warbler, or Summer Yellow-Bird

Length: About 5 inches.

Male: Olive-green above, bright yellow below; *breast streaked with brown*; wings edged with yellow; tail dark brown, with yellow on inner web; *no black on head, throat, wings*, or *tail*; bill slender.

Female: Similar; with fainter streaks on breast, or an unstreaked breast.

Song: A sweet *chee-chee-chee-chee-chee'-a-wee?*

Habitat: Orchards, gardens, and shade trees, rather than woods.

Nest: A beautiful cup lined with felt. This bird's nest has been recorded as a favorite depository for cowbirds' eggs.

Range: North America. Breeds from northern Canadian and Alaskan tree-regions to southern Missouri and northern South Carolina; winters from Yucatan to Brazil and Peru.

The Yellow Warbler is one of the best known of its tribe. It is an attractive, lovable little bird, a useful destroyer of small insects that feed upon the leaves of trees, and a charming addition to any orchard or garden, as it flits among the trees like a ray of sunshine.

It is frequently confused with the goldfinch; but careful observation of markings, of flight, and of song will show decided differences. The goldfinch has a black crown, wings, and tail, an unstreaked breast, undulating flight, and a sustained song. This little olive and yellow bird has no black in its plumage; it makes short flights, and sings a simple strain. It is not a seed-eater, like the finches, but is insectivorous.

---

[148] From "Birds of New York," page 421.

Figure 54. Yellow warbler.

The Pine Warbler

Length: About 5½ inches.

Male: Upper parts olive-green with a grayish tinge; throat and breast yellow; sides streaked with gray; belly white; wings and tail brownish-gray; wings with two whitish bars; *outer tail-feathers tipped with white on inner web.*

Female: Similar to male, but browner above and duller underneath.

Notes: "Its alarm note is a sharp chirp, its other notes are few and weak."

Song:

> "The song is one of the most soothing sounds of the pine-woods. It has in it the same dreamy drowsiness that characterizes the note of the Black-throated Green Warbler, but is otherwise entirely different in tone and quality, being composed of a series of short, soft, whistling notes, run together in a continuous trill. It resembles, in a way, the song of the Chipping Sparrow, except that it is softer and more musical."[149]

Habitat:

> "Pine woods and groves; it seems to prefer the pitch pines, and is one of the few birds that habitually live and breed in woods of this character, like those of Cape Cod. It has been called the Pine-creeping Warbler, from its habit of creeping along the branches, and occasionally up and around the trunks of pines."[149]

Range: Eastern North America. It is abundant in the South where pine forests are common. It is found in 270 southern Canada, northern and eastern United States, in such pine-regions as Michigan and New Jersey.

The Maryland Yellow-Throat

Length: About 5¼ inches.

Male: Olive-green above, brightest on rump and tail; yellow underneath, with gray sides; *a broad band of black bordered at the back*

---

[149] From "Useful Birds and Their Protection," E. H. Forbush.

with gray extends across the face in the form of a mask. The young males lack the conspicuous mask.

Female: Similar to male, but without a mask.

Note: A sharp call-note chick, frequently repeated.

Song: *Witch'-e-tee'-o, witch'-e-tee'-o.* Writers interpret the song in various ways. Mr. Forbush's *sich'-a-wiggle, sich'-a-wiggle, sich'-a-wiggle*, is an excellent rendering. The song varies with individuals, but is phrased and accented similarly.

Habitat: Roadside thickets, especially near water.

Range: Eastern North America. It breeds from North Dakota eastward to southeastern Canada, and south to central Texas, the northern part of the Gulf States and Virginia; winters from North Carolina and Louisiana to Florida, the Bahamas, Cuba, Guatemala, and Costa Rica.

The Maryland Yellow-throat is a delightful summer visitor. Trim, dainty, exquisitely colored, lithe, and full of song, he is a charming part of the thickets of roadsides and streams.

The Maryland Yellow-Throat

A host of warblers northward come in May,
And linger with us only one brief day;
You, yellow-throated songster, love to stay.
We glimpse your dainty coat of olive-green,
Your breast and throat of shimmering yellow sheen
And mask of black, where ferns and bushes lean
O'er sparkling streamlets, rimmed with many a reed,
And hung with brilliant golden jewel-weed.
Midst feathery spikes of meadow-sweet you speed.
Your brooding mate you watch, as to and fro
You flit; and while the summer breezes blow
You sing your *Witch-i-tee'-o, witch-i-tee'-o*.

Figure 55. Maryland yellow-throat.

The Hooded Warbler

Length: About 5½ inches.

Male: Forehead, cheeks, breast, and belly yellow; *back* of *crown* and *throat black*, the two dark areas *united by a black line*; *mask yellow*; back and rump olive; wings and tail a dark grayish-olive; the *outer tail-feathers largely white* on their *inner webs.*

Female: Similar to male, but without the black hood; dark edge to crown; breast faintly washed with black.

Song: E. H. Eaton in his "*Birds of New York*" writes: "The song of this warbler is one of the few which the author can hear with perfect distinctness and enjoy." He adds that it is described by Langille as follows: "*che-reek, che-reek, che-reek, chi-de-ee*, the first three with a loud, bell-like ring, the rest much accelerated with a falling inflection."

Habitat: Trees of deep woods.

Range: Eastern United States, west to the Plains, north and east to southern Michigan and Ontario, western and southeastern New York, and southern New England; in winter, West Indies, eastern Mexico, Central America, and Panama.

This warbler looks as though it had nearly divided a large hood,—had slipped one half of it back on its head like a calash, and allowed the other half to remain under its chin. It is easy to identify by its appearance and its song, and its habit of living in the lower parts of trees.

Eaton says:

> "The nest of the Hooded Warbler is usually placed in a low sapling or bush from 1 to 3 feet from the ground. In my experience it is the easiest of all the warbler nests to find. Wherever I have noticed a Hooded warbler singing in a patch of woodland, I have been very successful in locating the nest by placing my eye close to the ground and looking through the shrubbery from below the cover of the undergrowth. Then the nest will almost surely be seen if one is within a few rods, appearing like a bunch of leaves a short distance above the ground."

Wilson's Warbler

Length: About 5 inches.

Male: Olive-green above, except for a *black crown, outlined with yellow in front and at the sides of crown*; under parts yellow, except for a grayish tinge at the sides; wings and tail *without white bars* and *patches*.

Female: Similar to male, but without a clearly defined black cap.

Song: A loud, sweet trill, containing variations.

Habitat: Low thickets, usually at the edges of woods, rather than in treetops.

Range: Eastern North America. Breeds in the tree-regions of northern Canada south to southern Saskatchewan, northern Minnesota, central Ontario, New Hampshire, Maine, and Nova Scotia; winters in eastern 273
Central America; migrates through the Alleghanies; practically unknown from Virginia to Louisiana.

This attractive little warbler with its black cap might easily be confused with the goldfinch by a beginner in bird-study. The olive-green back, wings, and tail differentiate it. Unlike the goldfinch, it is not a resident, but a traveler to northern forests where it breeds. It journeys enormous distances.

"It appears very irregularly, some years in great abundance and some seasons not at all."[150]

The Black-Throated Green Warbler

Length: About 5 inches.

Male: Olive-green above; dull black patch below eye, *encircled with a broad rim of yellow*; *throat and breast black*, becoming yellowish-white on the belly; sides streaked with black; wings with two whitish bars; *tail with outer feathers largely white*.

Female: Similar to male; black of throat and breast mottled with yellow, streaks on sides less conspicuous.

Song: An insect-like trill, *zee-zee? ze-ze-zee?*

Habitat: Coniferous woods preferably.

Range: North America, from central Canada to northern Ohio and Long Island and in the Alleghany Mts., to Georgia and South Carolina; winters from Mexico to Panama.

For three summers I heard the persistent buzzing of this little Black-throat in the Maine woods before I was able to catch more than a fleeting

[150] "Birds of New York," Eaton.

glimpse of him. He is very shy and elusive. An opportunity to see this 274
beautiful little jeweled bird at close range is an event to bird-lovers. He is an industrious gleaner of small insects from dark pine and spruce forests.

The Canadian Warbler

Length: About 5½ inches.

Male: Gray above without white wing-bars or spots on tail; crown with fine black spots; *eye-ring*, and *line from bill to eye-ring bright yellow*; under parts bright yellow; *short black streaks extending across the entire breast*; white under tail.

Female: Similar to male, with fainter streaks on breast.

Song: A rapid and clear warble, more easily recognized than that of some warblers.

Habitat:

> "The Canadian Warbler during the migration season is found about our door-yard shrubbery, and the thickets on the edges of streams and woodland.... In the nesting season we must seek for it in cooler gullies or in damp, cool woodlands of deciduous or mixed growth."[151]

Range: Eastern North America. Breeds from south-central Canada to central Minnesota, Michigan, New York, and Massachusetts, and along the Alleghanies to North Carolina and Tennessee; winters in Ecuador and Peru.

The Yellow-Breasted Chat

Length: About 7½ inches; the largest of the warblers.

Male and Female: Olive-green above; bright yellow throat and breast; belly white; *broad white streak extending from bill above eye; white*

[151] From Eaton's "Birds of New York."

crescent beneath eye; 275 white streak at each side of throat, separating the olive-green and yellow areas.

Song: A medley impossible to describe, full of chucks and gurgles—a strange mixture of sounds. As a singer, the chat is in a class by himself; he is very different from the other warblers.

Habitat: Thickets and bushy pastures.

Range: Eastern United States; winters from Vera Cruz to Panama.

The following statements regarding the Chat are taken from Eaton's "Birds of New York":

> "The Chat is not a bird of the dense woodland or of open situations, but is confined to thick coverts of shrubs, vines, and young saplings, preferring a denser covert than even the Chestnut-sided warbler and the Catbird. It is rarely seen far from such situations....
>
> "Though the Chat is so averse to being seen, he will sometimes be found even within the limits of our villages and cities where suitable thickets of considerable extent are found and his loud song is frequently heard from village streets and sidewalks."

The Yellow Palm Warbler

Length: About 5½ inches.

Male and Female: *Crown chestnut; line over the eye and ring around the eye yellow*; upper parts olive-green, browner on the back; *under parts bright yellow, with streaks of brown on throat, breast, and sides*; wings sometimes edged with brown; tail edged with olive-green; outer tail-feathers with white spots on inner webs near tips.

Song: Two songs, one "thinner" and more rapid than the other.

Habitat: Fields and roadsides; feeds chiefly on the ground and among low bushes.

Range: Atlantic Slope of North America. Breeds in southeastern Canada and Maine; winters from Louisiana to northern Florida; casually to North Carolina and Pennsylvania. The palm warbler is the western species, an inhabitant of the Mississippi Valley and the region eastward. It is very common in Florida, where it may be discovered in company with yellow Palm Warblers.

This lively little warbler, with its nervous habit of tipping up its tail incessantly like a spotted sandpiper, resembles its near relative the yellow warbler in a few respects. The olive-green upper parts and yellow breast streaked with brown are points of resemblances, but the chestnut crown and yellow line over the eye are differences. Neither yellow warblers nor yellow palm warblers are dwellers in the woods, but prefer to live near the haunts of man. Yellow warblers are seen in trees and bushes, while the palm warblers are found by roadsides, often on the ground in the stubble of pastures, out in the open. While subdued in color and therefore inconspicuous, they are readily identified by the habit of moving their tails.

GROUP SIX—BLUE OR BLUE AND YELLOW WARBLERS

The Cerulean Warbler

Length: About 4½ inches; one of the smaller warblers.

Male: Upper parts bright blue; head and back streaked with black; light streak above eye; white throat, breast, 277 and belly, with a bluish-black line that extends across the breast and down each side; wings with two broad white bars; inner webs of all except the middle tail-feathers with small white patches near tips.

Female: Bluish-olive above, under parts pale yellow; light streak over eye; wings with white bars; tail-feathers with white tips.

Song: Mr. Stone describes the song of the Cerulean warbler as "an almost continuous '*zwee-zwee, zwee, wee-ee*' during the nesting season."[152]

Habitat: "They are numerous in the maple woods on the hillsides overlooking the swamp, as well as in the swamp itself," writes Mr. Stone.[152]

[152] From Eaton's "Birds of New York."

Range: Eastern North America. Breeds mainly from southeastern Nebraska, Minnesota, southern Michigan and Ontario, western New York, Pennsylvania, and West Virginia, south to Texas, Louisiana, and Alabama.

The Black-Throated Blue Warbler

Length: About 5¼ inches.

Male: Upper parts a *dull grayish-blue*, darker on the back, black bordering crown above the eye; *cheeks, throat*, and *upper breast black*; *belly white*; *sides black and white*; wings black, edged with blue, and with white next to body; *a white patch* on wing; tail bluish-black, *outer feathers largely white*.

Female: Very different from male; olive-green above, yellowish-white underneath; light streak over eyes; *white patch* near the base of the primary quills; tail bluish, with much less white than on males.

Song: "His song, though very versatile, is among the thinnest and most non-melodious of the family."[152]

Habitat: "Black-throated blue warblers prefer clearings amidst hemlock woods or along hemlock-clad gully banks where there are dense underbrush, bushes, and stump sprouts bearing multitudes of large leaves."[152]

Range: Eastern North America from Hudson Bay and Newfoundland south to the Northern States, and in the highlands and mountains to Massachusetts, Connecticut, Pennsylvania, and Maryland.

The Black-throated Blue Warbler, though not so brilliantly colored as many members of the family, is one of the neatest and best-groomed of all the warblers. As he flies from bough to bough or bush to bush he displays to fine advantage the clear black and white coloration, the white spots on the wings and tail flashing like the wings of a butterfly. He carries his wings and tail partially spread somewhat in the manner of the Redstart.... The male is not so nervously active as many other warblers....

> "This warbler's nest often contains an egg of the Cowbird. The nests are variously attached to slender scrubby bushes, 8 to 30 inches up, usually very close to old trails or old wood roads.... A constant characteristic of this warbler's nest is the decoration of decayed, spongy pieces of light colored wood fastened to the outside."[152]

The Blue-Winged Warbler

Length: About 5 inches.

Male: Crown and under parts bright yellow; a black line through the eye; back olive-green, yellower at the rump; *wings bluish-gray*, edged with olive and white; 279 *two broad yellowish-white wing-bars*; tail bluish-gray, with white patches of different sizes on outer feathers.

Female: Similar to male, but with less yellow on head,—on forehead and not on crown.

Song: "The song is insignificant, a wheezy performance of notes resembling the syllables '*swee-e-e-e-e, chee-chee-chee-chee*,' the first inhaled, the second exhaled."[152]

Habitat: "The Blue-winged warbler frequents swampy thickets but is sometimes found among the scrubby second growth of the hillsides and the undergrowth of dense woods."[152]

Range: Breeds in eastern North America from southeastern Minnesota, southern Michigan, western New York, Connecticut, and Massachusetts, southward to Missouri, Kentucky, Maryland, and Delaware; winters from southern Mexico to Colombia.

> "The Blue-winged Warbler is deliberate in its movements as compared with other warblers, acting more like a vireo than a member of its family.
>
> The nesting site of this warbler is on the ground in a bunch of herbs or at the foot of a small bush. The nest is surrounded by grass, weeds, ferns, or vines, which screen it effectively from view."[152]

The Golden-Winged Warbler

Length: A little over 5 inches.

Male: *Crown bright yellow*; white line over eye, *broad black line extending through eye*; *black throat bordered with white*; wings bluish-gray, with a *large, bright yellow patch*; upper parts, bluish-gray; under parts, 280 except throat, grayish-white; tail bluish-gray, with outer feathers nearly all white on their inner webs.

Female: Similar to male, but duller; cheeks and throat *dark gray* instead of black.

Song: "Its song is a 'lazy *zee-zee-zee*.' It has also an insect-like call-note, and a sharp *chip* alarm-note like that of the chipping sparrow.... The song when near at hand sounds like the syllables *zee-u-ee', zee-u-ee', zee-u-ee'*."[153]

Habitat: The beautiful little Golden-winged Warbler may be found in deciduous forests, especially among elm and birch trees, and has a habit of seeking the ends of branches for its food.

Range: Eastern North America. Breeds from central Minnesota, southern Ontario, and Massachusetts, to southern Iowa, northern Illinois, Indiana, and New Jersey, and northern Georgia; winters from Guatemala to Colombia; very rare in Florida and southern Georgia, and west of the Mississippi.

The Northern Parula Warbler

Length: A little less than 5 inches.

Male: Grayish-blue above, with a *bright olive-yellow patch in the middle of the back*; *yellow throat and breast, with a dark bluish or reddish-brown band across the breast*; belly white; sides sometimes reddish-brown; two white wing-bars; tail gray, edged with blue, with white spots near tips of inner webs.

---

[153] From Eaton's "Birds of New York."

Female: Similar to male, except that the reddish-brown markings and band across the breast are less distinct or wanting.

Song: A "buzzing" song rather evenly accented.

Habitat: "During the migration season, the Parula Warbler may be found among the foliage of our shade trees 281 and orchards, being most common about the time of the apple-blossoms. As soon as he reaches his summer home, however, he is practically confined to swamps ... preferring, during the nesting season evergreen trees, although occasionally found in mixed groves where deciduous trees predominate."[154] He lives in localities where he can find the *Usnea* moss, in which he loves to build his nest. Look for him along streams or near swamps where this moss hangs from the trees.

Range: From eastern Nebraska and Minnesota, central Ontario, Anticosti and Cape Breton Islands, south to Texas, Louisiana, Alabama, Virginia, and Maryland; winters probably in the Bahamas and from Vera Cruz to Nicaragua.

The southern species or parula warbler, differs slightly from his northern relative; his throat is yellower and his breast-band is less distinct. He lives in the southeastern United States, and is common where there are cypresses hung with moss. He is very active; he reminds one of the kinglet and the chickadee as he hangs head downward from a spray, seeking the tiny insects that he likes to eat.

Group Seven—Warblers with Reddish-Brown Markings

The Bay-Breasted Warbler

Length: A little over 5½ inches.

Male: *Forehead* and *cheeks black*, giving the effect of a *black mask*; *crown, nape, throat, upper breast*, and *sides* a beautiful chestnut-red; a

[154] From Eaton's "Birds of New York."

patch of buff at each side of the neck; lower breast and belly buff; back brownish-gray, 282 with black streaks; wings and tail brownish-gray; two broad white wing-bars; tail with white spots near tip of outer feathers.

Female: Upper parts grayish-brown, streaked with black; under parts buff, breast and sides washed with reddish-brown; crown brownish; two white wing-bars.

Song: "A monotonous, lisping song, with perhaps a few more musical, ringing notes."[155]

Habitat:

> "The Bay-breasted warbler usually frequents the tops of trees during migration, being especially fond of chestnuts, oaks, and hickories just as the leaves are bursting. It is also found in orchards and about the shade trees of streets and parks as well as in the midst of woodlands.... It prefers the upper portions of trees except in cold or stormy weather when it descends and feeds among the underbrush."[156]

William Brewster says that they live in dense woods, especially among the pines and other cone-bearing trees.

Range: Eastern North America. Breeds in north-central and southern Canada, northern Maine, and mountains of New Hampshire; winters in Panama and Colombia; irregular on the Atlantic slope and south of Virginia. One of our less common warblers.

The Chestnut-Sided Warbler

Length: About 5 inches.

Male: Crown *yellow*, bordered with black; back gray, streaked with black and yellow; *ear-patch and under parts white*; black line extending from bill meets *broad chestnut streak* which runs down the side of the

[155] James P. Chapin.
[156] From Eaton's "Birds of New York."

body; wings with two broad yellowish-white wing-bars; tail black, outer feathers with large white spots varying in size.

Female: Somewhat like male, but duller; the colors are less sharply contrasted.

Song: In the spring a loud warble, not unlike that of the yellow warbler; in the summer, a weaker trill.[157]

Habitat: Thickets, bushy roadsides, edges of woods, open woodlands.

Range: Eastern North America from central Canada to eastern Nebraska, northern Ohio, New Jersey, Rhode Island, and Massachusetts, and in the Alleghany Mts. to Tennessee and North Carolina.

The male Chestnut-sided Warbler is very easily identified; its sharp contrasts in coloring make it conspicuous. While the bay-breasted warbler also has chestnut sides, it differs in having the color extend to the breast and throat, instead of bordering the white under parts.

The dainty little chestnut-sided warbler is rather commoner than some species. Dr. F. H. Herrick in his book, "The Home Life of Wild Birds," tells of taming a female. She ate from his hand and allowed him to stroke her as she sat on her nest.

THE THRUSHES

Thrush Family—Turdidæ

Six members of the Thrush Family are more or less common in the eastern United States: the Robin, the Bluebird, the Wood Thrush, the Hermit Thrush, the Olive-backed Thrush, and the Veery. The Gray-cheeked and Bicknell's thrushes are not so widely known. The Russet-backed Thrush is the western representative of the Olive-back.

The Oven-bird, or Golden-crowned Thrush, and the Water-thrushes are not thrushes at all, but warblers, though they resemble thrushes in having brown backs and light spotted breasts, and in being dwellers of the woods.

---

[157] "Useful Birds and Their Protection," by E. H. Forbush, page 193.

The Brown Thrasher, sometimes wrongly called the Brown Thrush, also has points of resemblance—a speckled breast and bright brown back—but he is one of the Mimidæ or Mockingbird Family.

The breasts of young robins and the backs of baby bluebirds are spotted, showing their family relationship. Both robins and bluebirds have voices that possess a quality for which our thrushes are noted. I have heard the English thrush, famed in poetry. I consider its song inferior in quality of tone to those of our wood and hermit-thrushes, and veery; it strongly resembles that of our thrasher.

The true thrushes of our woods have backs of leaf-brown, varying in hue from bright russet to dull olive. Their breasts are white or buff, streaked or spotted; their tails are short; their eyes, large and lustrous. Their movements are quick, yet graceful. Their demeanor is gentle, though I have seen them strongly aroused when nest or young was disturbed.

The Wood Thrush

The Wood Thrush is the best known of these thrushes. It may be identified by *its large size* (a little over 8 inches); by its *bright brown head*, dull brown back, wings, and tail; white under parts that are *heavily spotted, especially on the breast and sides*; and by distinct streaks below the eyes.

Note: Its call-note is a sharp *pit*; its song a series of sweet cadences beginning with the liquid syllables *ah-oh-ee*?

Song: Four phrases often constitute the song, between which a soft purring sound is frequently heard, if one is near the singer.

Habitat: Wood thrushes may be found in open groves, parks, and wooded pastures, on large estates, and along secluded roads. They are rarely found near farm-buildings, but occasionally live in gardens and orchards.

A pair of thrushes once nested in a tree on a slope just back of a house where I chanced to be a guest. The mother-bird had begun her brooding, when carpenters arrived to build some steps near her chosen home. Frightened, she fled, and remained away for a time. Finally mother-love overcame her fears and she returned. The workmen were asked to do her no harm; they became interested in her, and she trustful of them. She let them approach within a few feet of her nest. We saw the shy wood-bird, serene and unafraid, raise her brood in the midst of noisy hammering, with friendly companionship close at hand.

Figure 56. Wood thrush.

The Olive-Backed Thrush

The Olive-backed Thrush is about an inch smaller than the wood thrush (7 inches), and is uniformly *olive-brown above*. Its *breast, throat, cheeks*, and *eye-ring* are *buff*; its *sides gray*. The breast, sides of the throat, and cheeks are spotted with black.

Note: Its call-note is *puck*;

Song: its song pleasing, with a phrasing that reminds one of the hermit thrush, but it is louder and less deliberate, and lacks, also, the hermit's liquid sweetness. The olive-back has a habit of singing from the pointed top of a tall spruce; nearby, on a neighboring treetop, an olive-sided flycatcher may utter its *Peep here*, or a hermit may sing in the grove below.

Habitat: The olive-back lives in woods, rather than close to the haunts of man; it prefers to be near streams and swampy places, as does the western russet-back thrush, a bird very similar in appearance and habits.

Range: The olive-back breeds in Canada and northern United States, and winters from Mexico to South America.

The Gray-Cheeked Thrush and Bicknell's Thrushes

Gray-Cheeked Thrush: "The Gray-cheeked Thrush is found in migration over all the Eastern States, but breeds farther north, beyond our limits.

Bicknell's Thrush:

> "Bicknell's Thrush, a closely related form, while having somewhat the same general range, breeds farther south and nests in the mountains of northern New York and New England. The species does not seem to be very abundant anywhere."[158]

---

[158] Bulletin 280, U. S. Department of Agriculture, Biological Survey.

Their resemblance to each other and to the olive-back makes them difficult to identify. The absence of buff from the head differentiates them from the latter species, which is a difference not readily observed except by experienced ornithologists. Bicknell's thrush is smaller than the gray-cheeked thrush.

The Veery

The Veery or Wilson's Thrush is slightly smaller than the wood thrush (7½ inches), and is a *lighter and more uniform brown above*. It has a whitish throat and belly, and *grayish sides*. The breast and sides of the throat are a soft buff, with faint spots of brown. Its light brown upper parts and its less conspicuous markings distinguish the veery from other thrushes.

Note: Its call-note is a whistled *whee'-u*,—loud, clear, and uttered frequently.

Song: The song is inexpressibly beautiful,—like organ-chords, or those that fill the Baptistery of Pisa when the Italian guide blends tones for the delight of listeners. A veery's song cannot be described; the *whee'-u* may reveal the singer's whereabouts, and aid in identification. This bird has brought me pleasure many times, for it forms one of the chorus that sing their matin- and even-songs in a spruce grove across the road from our cottage in Maine. Still other veeries chant with hermit thrushes in more distant woods.

It recalls, also, memories of deep Adirondack woods near Seventh Lake, where we heard veeries and wood thrushes sing antiphonally at sunset.

Range: This thrush is abundant in the eastern United States during its migration, while on the way to its nesting place in our Northern States, to New England, and Canada. It winters in South America.

It seems to bear a charmed life. It does no harm and receives none; it is a favorite wherever its voice is heard.

The Hermit Thrush

The Hermit Thrush may be described in superlatives. Of the four commoner thrushes, it comes earliest (in March or early April) on its way to its haunts in northern woods, remains longest (till October or November), and is considered by many to be the finest singer of a highly gifted family.

It is so very shy that it is rarely seen and yet, during migration time, I once discovered a solitary hermit in a tree on a vacant lot only a few blocks from the business center of Cleveland. Because we sit quietly for hours at a time in the Maine woods, we have been vouchsafed many glimpses of its olive-brown back, *its reddish-brown tail* (the mark of identification), and its rather thickly spotted white and brown breast. We have noticed its habit of raising its tail as it alighted; we have heard its call-note *chuck*.

Moore's Rock, Castine, Maine, commands an enchanting view of Penobscot Bay, of distant hills, and of spruce woods that are tenanted by veeries, olive-backed, and hermit thrushes. There we make frequent pilgrimages, to hear them sing at sunset.

Beneath glowing skies and in the silence, the hermit raises his exquisitely modulated voice in a strain of ethereal beauty; pauses, then in a higher key, repeats it; a third time, with still loftier elevation of tone, he sings,—and sings again.

More than once at twilight, a white fog has moved in from the bay and enveloped us as we listened. The voices of these thrushes, proceeding from the sea of mist, have seemed more like those of spirits from another world than of birds—unspeakably uplifting and full of significance.

Chapter 9

AFTERWORD

The great psychologist, William James, preached the doctrine that it was immoral to have emotions that did not bear fruit in action,—a doctrine that many educators and teachers are putting into practice nowadays.

Music, art, noble architecture, poetry, fine prose, the drama, and the beauties of nature, all of which arouse the emotion of joy and minister to our higher natures, were formerly sought as means of self-development or culture—one of the great ends to be attained in life. Excessive cultivation of one's self is now regarded by broad-minded people as a refined form of selfishness (often intellectual snobbishness), unless with it there exists a sense of responsibility and an attempt to assist in making possible by some form of activity a more nearly universal sharing of these pure forms of pleasure.

The conservation of forests, the preservation of scenic wonders, of wild flowers, of native animals and birds for the enjoyment of all, has become the aim of a great movement throughout the country. It is well known that the fine balance of nature is maintained by birds, and that upon them depend in large measure the preservation of forests, parks, gardens, orchards, and farms.

As they are so truly our benefactors and furnish us with so much genuine enjoyment and absorbing interest, we are under obligation to repay their services to us by some form of service to them, which will minister also to the well-being of our communities. The formation of 291 Audubon Societies, the spreading of knowledge by means of bird-books, illustrated bird-lectures, and the invaluable bulletins easily obtainable at the Government Printing Office, Washington; the erection of bird-houses and baths, and of feeding-tables for the winter; the furthering of wise legislation regarding bird-protection and the supplying of bird-wardens in some localities to help carry out the laws; intelligent and humane regulations to prevent the depredations of cats; the creation, wherever possible, of bird-sanctuaries and preserves, and the planting of trees and shrubs which will attract birds are a few of the ways in which we may make practical our interest in birds and add to the well-being of our land.

INDEX

B

C

N

O

P

Q

R

S

T

V

W

Y